Society, Action and Space

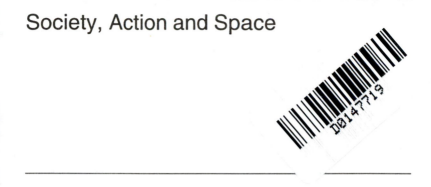

This book questions the relevance of space for the social world. It focuses upon issues which are at the centre of important debates in human/social geography. Over recent years one of the most significant developments in social analysis has been the increasing interchange between geographers, sociologists, anthropologists and social philosophers concerning 'the spatial'. This debate involves the work of Giddens, Foucault, Bourdieu, Lefebvre, Harvey, Gregory and Soja and many others. As a result of these new developments a whole series of new forms of empirical work as well as theoretical innovations have come into being. Spatial considerations are no longer confined to the area of geography. Rather they are now seen as fundamental to all forms of social theorising, particularly in social conditions where spatially very distant events now affect everyone: under conditions of late modernity and globalization.

The distinctiveness of this book is that it connects discussions in the philosophy of social science with theories of action which have direct relevance to concepts of space. The book provides a discussion of Popper's critical rationalism, linking this with ideas drawn from phenomenology. This epistemological debate is linked with the sociological action theories of Pareto, Weber, Parsons and Schutz. The book closes with an evaluation of how 'the spatial' can be systematically integrated into action theory.

The book offers exciting new directions for geography and sociology as 'spatial sciences'. However, it insists that 'space' in itself does not cause or explain anything. It only becomes meaningful in the context of action theory.

Benno Werlen lectures in Geography at the University of Zurich.

Society, Action and Space

An alternative human geography

Benno Werlen

Translated by Gayna Walls
Edited by Teresa Brennan and Benno Werlen

London and New York

First published in German in 1988 as
Gesellschaft, Handlung und Raum by
Franz Steiner, Stuttgart

First published in 1993
by Routledge
11 New Fetter Lane, London EC4P 4EE

Simultaneously published in the USA and Canada
by Routledge
a division of Routledge, Chapman and Hall Inc.
29 West 35th Street, New York, NY 10001

© 1993 Benno Werlen

Typeset by LaserScript Limited, Mitcham, Surrey
Printed and bound in Great Britain by
Biddles Ltd, Guildford and King's Lynn

British Library Cataloguing in Publication Data
A catalogue record for this book is available from
the British Library.

Library of Congress Cataloging in Publication Data
Werlen, Benno.
  [Gesellschaft, Handlung und Raum. English]
  Society, action and space: an alternative human geography/
  Benno Werlen:
  translated by Gayna Walls. – 1st ed.
  p. cm.
  Translation of: Gesellschaft, Handlung und Raum.
  Includes bibliographical references and index.
  1. Human geography – Methodology. 2. Human geography –
  Philosophy.
  GF21.W4713   1992
  304.2'3 – dc20   92-6606
                                                       CIP

ISBN 0–415–06965–3
      0–415–06966–1 (pbk)

Contents

Author's foreword to the English edition

If a text is translated, it is not only put into another language but also into another context, another tradition of thinking. For this reason a simple translation is not sufficient. Rather what we need is the transposition of a line of argument from one cultural space into another. But this act of translocation is not simply linguistic. So in fact I should have rewritten this text for its new purpose. This would have had the advantage of ridding it of all aspects I now see differently. But then the chance to animate an intercultural discussion between social scientists and geographers in the German and Anglo-Saxon traditions might be missed. The present form of this book is ultimately a compromise between these different positions. But before talking about the changes I made to this English edition let me briefly say something about the book's origin.

When I started to work on *Gesellschaft, Handlung und Raum* in 1980, the situation in social theory and human geography was quite different from today. The key debates, if they have not moved on, have certainly changed in focus. In the social sciences in 1980 structuralism, Marxism and Parsonian sociology dominated theoretical discourse. It was obvious to me then that we needed a theoretical framework that would enable us to give a more accurate account of actual social processes, and of the role of the subject in them. We still need that account, although today there is more awareness of questions concerning 'the return of the subject'. Despite this awareness however, a theory of how the subject can return is lacking. I realized that my original argument bore exactly on this question. I rewrote the English edition with this in mind.

The different national contexts also need to be taken into account. In the German-speaking world in 1980 the geographical debate was still dominated by the spatial scientific approach. The most famous exponent, Dietrich Bartels, elaborated a much more sophisticated version, than the Anglo-American geographers like Bunge, Berry and others. The latter did not go beyond a neo-positivist conception of science. Bartels was, however, looking for more than that. He was searching for a critical–rational foundation of geography as a science of the spatial combined with general system theory. But because he overemphasized space, his conception could not satisfactorily demonstrate how to analyse the intersection of subjectivity, society and space.

After I finished a longer project on functionalism in social sciences and human geography I saw the main problem of human or 'social' geography – as we prefer to say in the German-speaking world – in this way: geographers take 'space' or 'landscape' as an object (sometimes even as a causal factor determining social processes). Sometimes they try to think the whole of human existence in spatial categories.

I would add that in the German-speaking context 'social geography' has a much wider connotation than in the Anglo-Saxon tradition. Ever since Hans Bobek and Wolfgang Hartke started to develop research concepts out of this geographical subdiscipline in the late forties and early fifties, all theoretical debates on human geography focused on social geography. This term replaces 'cultural geography' and also encompasses, to a certain extent, economic geography.

From the outset, all these theoretical debates were unable to specify the relationship between 'subjectivity', 'society' and 'space'. In the early period they were dominated by the idea of landscape as the core research object of geography. Bobek's aim was to classify socio-geographical groups as the main producers of landscape shapes. He wanted to find socially and spatially delimited units of populations to explain the shape of a given landscape. Hartke maintained this approach, not to explain landscape, but to explain the spatial divisions of societies. The concept of 'reach' is central. In Hartke's view, the spatial division of societies is the expression of the spatial reach of human activities directed by the same values and norms. Therefore the first aim of social geography should be to determine regions of human activities guided by the same norms and values. In Anglo-Saxon geography there is today, I think, a similar tendency. It finds its expression in the ambition to discover spatial delimitations to social classes.

In German-language geographical studies these starting points gave direction to subsequent research in the late sixties and early seventies. The so-called Munich School of social geography formulated a functionalist reconceptualization of Bobek's and Hartke's ideas. The Munich starting point, proposed by Karl Ruppert and Franz Schaffer, is based on the conviction that all human activities that matter for social geography are nothing more than the expression of seven basic needs. These are an extended version of the four key functions developed by Le Corbusier for functionalist architecture and urban planning: 'housing', 'labour', 'to provide', 'to educate', 'to recover', 'traffic' and 'living in community'. These categories were meant to provide a key to understanding landscapes, analysing the spatial division of societies and as the basis for urban and regional planning which would achieve a spatial equilibrium of functional facilities. Empirical research was therefore oriented towards the discovery of specific 'action spaces' for each function. In current Anglo-Saxon human geography the equivalent research concept is based on the idea of finding specific 'action spaces' for given social categories.

More or less at the same time, Bartels' social and economic geography concentrated on the discovery of laws in the spatial distribution of social facts ruling human activities. He suggested a three-stage research programme leading from the spatial localization of social facts to the formulation of regularities in the functioning of social and spatial systems. Here again the idea of a spatial equilibrium is of central importance. But, as with all other concepts in social geography, human activities and the social are not the research 'object': space is. Subjectivity, human activities and social facts have to be reduced to spatial categories.

People's activities are not just functions or the effects of a thing like 'space' nor are they just a response to the so-called spatial environment. My distinct impression is that every scientific account based on spatial fetishism or an overspatialization of the social world is leading to repressive discourses and politics. But the question remains: how can we provide an appropriate account of human subjective activities without neglecting the spatial dimension? It is obvious that the spatial dimension matters to social reality even if, as I shall argue, 'space' does not exist as an object or a causal force. But how can we deal with that on a theoretical level?

In social science there are four main directions of the thematization of human activities. There is the tradition, where, obviously, 'behaviour' is the central concept: every activity is governed by a stimulus. There is Marxism, concentrating on 'labour'/'production'. There is the sociological system theory based on the idea that society is constituted by a set of subsystems of different types of activities. Finally there is action theory, emphasizing the creative aspect of human activities by the concept of intentionality, denying causal determination in the natural scientific sense. Action theory's account of social processes is formal. For that reason this concept can be used for different topical fields from sociology to economy, law, psychology, social/ cultural anthropology, etc. It is not limited to one specific type of activity like 'labour' or to physiological responses like 'behaviour' and is not limited to mechanistic accounts of the social world in the way that system theory is. These are the main reasons why I was tempted to apply action theory to the research field of social geography.

In the early eighties Peter Sedlacek, a German geographer, produced a paper on 'social geography as normative action science'. This publication endorsed my reasoning. Sedlacek's theoretical approach is based on the constructionalist philosophy of the Erlangen School, represented by Mittelstrass, Lorenzen and Schwemmer. Their aim is to elaborate a method for the formulation of legitimated norms of action under conditions of democracy. Their final conclusions resonate with Habermas's concept of a domination-free discourse of mutual understanding. Researchers should therefore participate in social processes and help the actors, or subjects as we would say today, to find adequate means for (unproblematic) goals in respect to the existing norms. If there are conflicting aims, researchers should – in a domination-free discourse – help to formulate norms as standards for the decision between these aims and to assure the agreement of all the parties in a conflict. Sedlacek applied this method to problems of localization, i.e. on the level of urban and regional planning. Spatial patterns become in this view the results and the conditions of action.

This is certainly an important step away from traditional spatialism. But Sedlacek's concept of action is too narrow. This can be explained by the fact that he ignored sociological action theories, and was more interested in the field of economic than on broader social geography. In addition, Sedlacek's method starts from a rather naïve view of the social world; as I have indicated, it assumes a normative position, before the descriptive and explanatory job is done.

On the other hand – and this was the widespread opinion at that time – social scientists do not take 'space' into account at all. So the problem was how to find a way to consider the social and the spatial, without falling into a spatial fetishism, and without excluding the spatial dimension from social theory. For me, the turning point in this dilemma was, surprisingly, the three-world theories, elaborated in different ways by Popper and Schutz. Popper and Schutz have probably received more critical attention in the German-speaking world than they have elsewhere. My initial attitude to Popper was very critical. I intended to use him to show the shortcomings of a technocratic spatial approach and a restricting view of social sciences. When I actually read him, I discovered another Popper. Although Popper had been used to justify spatial fetishism, Popper's writings in fact provide no such justification.

I started to realize that we have to make a clear distinction between the material, the mental and the socio-cultural aspects of reality. Geographical spatialism deals with nothing but the material in the shape of a product of reification: the object called 'space'. Therefore explaining the socio-cultural by space is only a hidden form of vulgar materialism, hidden by an unacceptable operation of reification. In addition, trying to localize mental or socio-cultural facts by means of the geographical concept of space implies a very crude reduction. Human agency and every human action has a socio-cultural, subjective and material component and it is not just misleading but also inappropriate to take only one of these aspects into account for their explanation.

Schutz, like his mentor Husserl, had also been misread. The exponents of humanistic geography – just starting to take an important place in the theoretical debates in the early eighties – had tried to use Schutz's phenomenology to prop up behaviourism. These geographers went on talking about space in a quite traditional way even if they provided many important criticisms of the spatial approach. And this is in my opinion one of the reasons why they combined phenomenology with behaviourism. Just as behaviourists claim that the perceptions of material things are stimuli of our activities, so the 'phenomenological humanists' in geography claim that the spatial – or spatial informations – prompt and guide what we do. Yet, on reading Schutz I found that he himself was really interested in the subjective perspective based on action theory.

Popper and Schutz are by no means the only starting points for an alternative human geography in which the subject returns. There are other action theories in the classics of the social sciences. But what the classical sociologists, Weber, Pareto, Parsons and others, did not emphasize enough – except Schutz – is the central role of the human body in social life in general.

In this edition the analysis and criticism of the German tradition of social geography – the last chapter of the original publication – have not been translated. There I demonstrated the implications of the change of perspective from a space to an action-centred geographical research by reformulating Christaller's central place theory. The other important exision is a chapter on different logics of the explanation of actions in the objective and subjective research perspective; I do not think that these cuts affect the main line of the argument.

On the other hand, in the first chapter I have added a more general discussion on the status of space. And I completed Chapter 6 with a section on Bourdieu's concept of social space. This concept was published only after I finished the German version of this book and comes very close to my standpoint. I do hope that these changes will help to make my proposition of an alternative human geography more intelligible.

Zurich, January 1992

Preface

Anthony Giddens

This book will be of interest not only to geographers but to everyone working in social theory. 'Space' has sometimes been claimed as the distinctive province of geography, but in this book Benno Werlen demonstrates brilliantly that such a position is not plausible. 'Space' has no distinctive content and therefore cannot be the specific object of study of a particular discipline. This conclusion does not imply that spatial issues lose their interest; on the contrary, they become of core importance for the social sciences as a whole.

The concept of space, as Werlen indicates, is philosophically as problematic as that of time. Time, some have asserted, does not exist: the 'temporal' is no more than a way in which events are ordered. Something of a similar claim has been made about space by Leibniz and others. Without engaging with such philosophical issues in any detail, Werlen holds that a 'substantialist' interpretation of space is not defensible. Geography became widely understood as a 'spatial science' as a result of a wrong alley taken early on in its development: Kant's categorical notion of space became translated into an explanatory endeavour. The quest for 'spatial laws' was on. Such a quest, in Werlen's view, is a nonstarter; but this does not imply that geography cannot be explanatory. The explanatory content of geographical investigation can be redeemed if geography becomes an action-oriented enterprise. Much of Werlen's book is devoted to an exploration of the nature of action. Problematizing the notion of action, as he often emphasizes, is not to be equated with voluntarism or individualism. Moreover, since action presumes an agent, and agents are embodied, an approach based on the centrality of action has a definite 'materiality' to it. An action approach does not imply subjectivism.

In developing these themes, Werlen offers a rich and provocative analysis of various influential traditions of social thought. The works of Karl Popper, Alfred Schutz, Talcott Parsons and others are subjected to acute and thoroughgoing critical interpretation.

A stress upon action can help counter the behaviourism and objectivism characteristic of many areas of geography, but of course 'action' is a complex idea. Action presumes meaning and intentionality, even though what counts as action is not exhausted by what is done intentionally. Yet we must also have an account of the unintended consequences of action, a perennial feature of human social life. Popper's writings provide a major resource, Werlen proposes, for investigating these issues. Yet Popper's work has led to an elemental inconsistency. Popper quite correctly rejects empiricism and inductivism. According to a common misconception, Popper's critical rationalism advocates a unity of method between social and natural science: all generalizing sciences, whether of nature or of human activity, follow the same methodological precepts. In Werlen's view, there is a related error. Although the logic of falsification might be universal, the same does not hold for explanatory postulates. In developing this argument, Werlen provides an important clarification of Popper's methodological individualism. Methodological individualism, he shows, does not imply an ontological premise about individuals – it does not imply that 'only individuals exist'. An orientation to methodological individualism is compatible with recognizing the reality of social collectivities or institutions. The point is that collectivities never 'determine' what individuals do. This is only another way of emphasizing, therefore, the primacy of action as the basic preoccupation of social science.

Generalizations in the social sciences, Werlen says, are not 'laws': instead, they relate to common patterns of action, or 'mixes' of intended and unintended consequences of activity. Popper recognizes this point, in a certain sense at least, when he speaks of the significance of 'situational analysis' in the explication of human conduct. When subjected to scrutiny, however, situational analysis can be seen to diverge from the explanatory logic of natural science. The only 'causes' to be found in the social sciences are agents' reasons, and these cannot be assimilated to orthodox models of scientific causality. Werlen does not choose to follow the detour leading through the writings of Peter Winch, and their critical reception, but he explores in an acute way some of the fundamental questions of action, meaning and institutions which Winch and others brought to the fore.

Popper's theory of knowledge 'without a knowing subject' actually coincides with ideas coming from very different perspectives – those sometimes associated with 'subjectivism'. Among these perspectives is that of Schutz, whose views Werlen also analyses in some considerable detail. Schutz's ideas at first sight seem completely opposed to those of Popper. The 'knowing subject' is, after all, the prime point of reference for phenomenology. Social knowledge, however, is understood in Schutz's work in terms of the 'natural attitude', which is intrinsically intersubjective in character. Ontological individualism is discarded and action is again placed in the centre. Schutz's conception of explanation in social science, like that which Werlen imputes to Popper, distances itself from naturalistic versions of causality. The social world is constituted as meaningful in the everyday activities of social agents themselves. The explanatory tasks of social science are bound up with constructing 'second order' constructs with meaning which allow for the reinterpretation of actors' own meanings. Schutz's celebrated notion of 'adequacy' presumes that accounts offered by the social observer must be in principle intelligible to ordinary lay actors. Adequacy on the level of meaning has to be complemented by causal adequacy: a construct is causally adequate if the course of action which it projects conforms to the typical conditions under which a given act is undertaken.

In a masterly synthesis, Werlen draws out the common features of the views of Popper and Schutz. Popper and Schutz both suppose that social science should begin from the study of action, not structures. Each agrees that the social world is not organized in terms of any sort of causal determinism; both give a considerable role to ideal-typical concepts in their interpretations of human activity. Yet we cannot be content only with a synthesis. We must go beyond Popper and Schutz, Werlen argues: for while each correctly locates the problematic of social science in the notion of action, neither explore that notion in a fully satisfactory fashion. In furthering such a task, Werlen turns to Max Weber, Vilfredo Pareto and Parsons. The ideas of these authors, of course, themselves cannot be accepted in an uncritical vein. Appropriated selectively, however, they provide notions which can be used to flesh out an elaborated theory of action. All action occurs within 'frame of reference' which agents employ in reacting selectively to given situations, but which are also constantly reformed in the light of their experience of those situations.

What are the implications of all this for geography and for the problem of space? To answer this question, Werlen turns to a direct discussion of the concept of space. 'Spatial problems' in geography, consistent with Werlen's emphases, always refer to issues relating to action. Space cannot 'cause' or determine anything. To speak of 'space' in the context of geography is in fact to broach a series of concerns to do with the arrangement of persons and objects in the physical world. Location is only socially relevant – and this is crucial – when filtered through the frames of reference that orient individuals' conduct. In these terms, Werlen is able to effect a reconciliation between social geography and sociology. Each preserves a certain distinctiveness, yet each contributes directly to the other.

Werlen's arguments are well documented and persuasive. They sound the death-knell for the idea of geography as the 'science of the spatial', but at the same time open up new and remarkable fields of study. It is to be hoped that this book will have just as great an impact in the English-speaking world as it did in its German context of origin. The reader will find here a work of subtlety, originality and insight.

Acknowledgements

Many people have helped me, directly or indirectly, in the process of preparation and of writing the German and 'rewriting' the present edition. My first thanks are to Anthony Giddens, for his encouragement and intellectual counsel. He created a space for action which enabled me to finish the English version of this book. I am grateful to Jean-Luc Piveteau for guidance on the first draft. I would also like to thank the late Dietrich Bartels, Gerhard Hard, Otfried Höffe, Albert Leemann, Ed Swidersky, Jean Widmer, and many others in the German-speaking academic world for their helpful comments on the first edition. My gratitude is also due to the following people for conversations and corrections. Deirdre Boden, Roger Brunet, Tommy Carlstein, Stuart Corbridge, Derek Gregory, Torsten Hägerstrand, Günter Krebs, Yves Lacoste, Heidi Meyer, Gunnar Olson, Chris Philo, John Pickles, Claude Raffestin, Jean Bernard Racine, Dagmar Reichert, Andrea Scheller, Graham Smith, Nigel Thrift, and Peter Weichhart. Finally, I would like to thank Teresa Brennan, for her brilliant editing and revisions of Gayna Walls' careful translation, and her cat Ptolemy for sitting on the manuscript whenever I had reached a really critical point in thinking.

Chapter 1

Space and causality, or
Whatever happened to the subject?

In the past many geographers argued that the aim of geography was to study space. Today most geographers argue that the aim of geography is to analyse the significance of space for social processes. Either way, the evocation of 'space' is crucial in justifying and differentiating our subject. Now if this line of reasoning is a legitimate one, 'space' has to be both an object of research and a meaningful constituent of 'social processes', and processes can only be social if they involve human action at some point.

In philosophy the nature of 'space' has been the subject of extensive debate. The context is different, but the issues are geographically relevant. In the present debate in philosophy, there are three positions: an absolute or substantive one, a relational one and an epistemological one. Let me get straight to the philosophical points, before turning to geography. The three philosophical positions are illustrated by these examples.

The absolute/substantive idea:

> For the idea of extension that we conceive any given space to have is identical with the idea of corporeal substance
>
> (Descartes).

> Absolute space, in its own nature, without relation to anything external, remains similar and immovable. Absolute space is the sensorium of God
> (Newton).

The relational idea:

> I hold space to be [. . .] an order of co-existence. Space is nothing at all without bodies, but the possibility of placing them
>
> (Leibniz).

The epistemological idea:

> *Space is not an empirical concept (Begriff) which has been abstracted from outer experience. [. . .] Space is a necessary representation, and consequently is a priori*

(Kant).

The philosophical debate has led to the conclusion that there is no way that the substantive idea about space can be maintained. Without going into details of the debate, which are available else-where,[1] one important point should be singled out: if 'space' were an object, that is to say, an appropriate research object, then we should be able to indicate the place of space in the physical world. But this is impossible. Space does not exist as a material object, or as a (consistent) theoretical object.

How then did geography come to be defined as a 'spatial science'? We can trace this categorical mistake back to one of modern geography's founding fathers, Alfred Hettner's mis-understanding of Kant.[2] For Kant (1802) geography was a descrip-tive or taxonomic discipline,[3] rather than a science: it only had the status of a propaedeutic discipline. Kant used the word 'choro-graphic', meaning descriptive, to describe geography. Hettner (1927, 115f., 127ff.) transformed this into 'chorologic', which refers to explanation rather than description, and explanatory force was the cornerstone of Kant's definition of science. Hettner's 'mistake' made it possible to describe geography as a science: the science of space. Berry, Bunge, Bartels and others walked down to the end of this dark street. Bartels (1968, 1970), the most famous German geographer of recent decades, finally attempted to formu-late the aim of geography as the discovery of spatial laws. In this, he lapsed back into a substantive concept of space. He also attempted to justify his argument by reference to Popper's criteria of what constitutes a science.

Yet if we reject Hettner's misinterpretation and all that followed from it, this does not mean we have to accept Kant's definition of geography. I believe it is more than just a propaedeutic discipline, in that it has explanatory potential. But I am also convinced that to define and maintain the idea of geography as a social science, the central role accorded to space has to be replaced. It has to be replaced by 'action', as the key concept of geographical research. I shall return to why I prefer 'action' to 'social processes' (although they come to the same thing in the end). The immediate point is that space is neither an object nor an *a priori*, but a frame of reference for actions. Space is a frame of reference for the *material* aspects of social actions in the sense of a formal-classificatory concept. In contrast to Kant, I suggest the following definition of 'space':

> *'Space' is not an empirical but a formal and classificatory concept. It is a frame of reference for the physical components of actions and a grammalogue for problems and possibilities related to the performance of action in the physical world.*

The framework cannot be empirical because there is no such thing as 'space'. 'Space' is a formal frame of reference because it does not refer to any specific concept of material objects. It is 'classificatory' because it enables us to describe a certain order of material objects with respect to their specific dimensions. This conceptual framework enables us to take account of the material implications of social actions in relation to the physical world, and the corporeality of the actor, who now features in much social and contemporary theory under the name of the 'subject'. As an aside, I should add that the shift from the word 'actor' to the word 'subject' is also a shift from a term that embodies action, to one that suggests that human beings are subject to forces beyond their control. I would hope that new thinking about the implications of this shift may be prompted by this argument.

If we begin from the perspective of what, in this book, I shall term 'action theory' or 'action-oriented geography', and discard 'space' as a starting point in itself, we begin from a perspective that focuses on the embodied subject, the corporeality of the actor, in the context of specific socio-cultural, subjective and material conditions. We begin from a perspective that emphasizes subjective agency as the only source of action and hence of change, at the same time as it stresses that the social world shapes the social actions that produce it. But the fact that the social world is produced and reproduced by social actions means that it is these actions, rather than 'space', that is constitutive of that world. Any concept of space can only provide a pattern of reference by means of which problematic and/or relevant material entities that bear on actions can be reconstituted and localized. Given that the subject is embodied, these material patterns are of course significant in most actions. But since they are not the only significant factor in actions, actions cannot be explained by them. The socially constructed frame of actions is not a 'spatial' cause; it is the product of actions. This means that it is insufficient to proceed from the assertion that 'space' or even materiality already have a meaning 'in themselves', a meaning that is constitutive of social facts. Materiality only becomes meaningful in the performance of actions with certain intentions, and under certain social (and subjective) conditions.

All this runs counter to Henri Lefebvre's idea of space. To the question 'What exactly is the mode of existence of social relations? Substantiality? Naturality? Formal abstraction?', he replies, 'The social relations of production only have a social existence insofar as they exist spatially; they project themselves into a space, they inscribe themselves in a space while producing it.'[4] This answer is problematic in many respects.

First, the social content of social relations (of production) cannot be 'spatial' at all because of the immateriality of the social component. I shall explain what I mean by this. The social content of a social relation of production can be nothing else but a meaningful attribution to material facts by subjects with given aims in view at a given moment under certain socio-cultural conditions. This sociality as such does not exist in materiality. Otherwise every material object could have only one social meaning. But we all know that it is possible to attribute different meanings to the 'same' material fact, just as it is possible to attribute different meanings to the same text (cf. Derrida 1987). *Second*, Lefebvre's formulation implies a pre-existing substantialized 'space', in that he writes that 'social relations of production [. . .] project themselves into a space'. And *third*, a 'social relation of production' by itself can do nothing. It becomes socially relevant only in processes of action produced or reproduced by an actor.

Lefebvre's formulation involves a double reification: the reification of space, and the reification of 'relations of production'. The first leads to a reductive materialist view of the world: the 'social' only exists when it becomes spatial. Because only the material can be spatial the implication of this assertion is that the social only exists in the material and can be reproduced only by the material. The only possible conclusion is: the material is acting on the material and in this process the social is generated. Needless to say, despite Lefebvre's profoundly argued disagreements with them, this conclusion leaves him very close to the geo-determinists he argues against. Lefebvre does not solve geographers' problems.

Not only does he lapse into vulgar materialism; in defining 'space' as a research object, he is also essentialist. For he reifies 'space', effectively giving it a content with a determining power. In a similar manner, the idea of 'time' in history was reified, and misunderstood as a power in itself: the 'power of time'.

The main difficulty in current research strategies is that geographers try to localize immaterial social and psychological entities in the physical world. This localization is impossible because, as we shall see, these entities have different ontological status. In the physical world, only material entities can be localized. Immaterial entities cannot. The formal concept of space can localize material things, but its categories (longitude, latitude, etc.) are not adapted to other phenomena. If we accept that actions always have at least a socio-cultural, subjective and physical component, it should now be clear that spatial categories can only grasp the last of these components. Every attempt to grasp the immaterial socio-cultural and subjective worlds of intentions, norms, values, etc. in spatial terms is accordingly reductionist.

Were we to rely solely on the physical categories (to which space is assimilated) in accounting for the socio-cultural world, we would have to assume that the meaning of actions is causally determined by the biological body, or other material conditions of the situation of action. This reduction is not a fantasy; it figures directly in some of the theories I shall discuss, which do reduce the subjective and socio-cultural aspects of actions to a biologically determined cause. In this sense, so-called spatial arguments resonate with racist and sexist explanations in their representation and explanation of the social. They resonate because they use material, biological categories in accounting for socio-cultural facts. Ironically, this brings (vulgar) materialism, claiming to be Marxism, very close to behaviourism, and the homogenization that characterizes all totalizing or holistic explanations.

Homogenization is expressed in the form of generalized prejudices. A common form is the attribution of 'common characteristics' to, for example, Africans or Hispanics, which denies differences among different ethnic groups, and, by this denial, obliterates the singularity that white, privileged groups reserve thereby for themselves. Attributing difference in its own way denies difference, at the same time as it helps to perpetuate differential inequalities. The holism that characterizes the types of explanation mentioned above is expressed in the belief, when it is based not on social but on physical categories, that a collective can act 'in itself'. The most prevalent form of this approach is found in nationalist or regionalist arguments professing to represent the 'will' of, for instance, the Lithuanians or Slovenians. A regional category takes over entirely from that of subjective agency. The point is not that Slovenians do not feel themselves to be Slovenians, but that this (regional) categorical takeover denies them their decision-making power as subjective agents.

Instead of searching for descriptions and explanations of the social world in spatial categories, geographers should be able to provide explanations of the so-called spatial facts in categories of action. Or, more accurately, geographers should be able to give explanations of actions referring to the constraining and enabling aspects of socio-cultural, psychological, *and* material factors in terms of the conditions and consequences of actions.

This book has the subtitle: 'an alternative human geography'. This raises the obvious question: An alternative to what? As I have spent several paragraphs on a leading and valuable Marxist theorist (Lefebvre) I should make it plain that my main opponent is not the alternative Marxist geography propounded by thinkers like Harvey (1985, 1989), but the geo-determinist, behaviourist assumptions that have governed the subject of geography ever since it named itself a spatial science. It is the researchers so governed that still secure academic appointments and get the research grants (money) in spite of their intellectual bankruptcy, and this is why it is important to prove, in terms that their students, if not these researchers themselves, will recognize, that the grounds on which they stand are shaky, if not shifty. Marxist theories of geography are implicated in my critique insofar as they fall into the trap of thinking, together with the geo-determinists, that space is a cause. And into the related trap of dismissing subjective agency, the essence of any real action theory, as voluntarism or individualism. Moreover, my emphasis on the constraints on action and the shaping of socio-cultural values reflects my complete agreement with the assumption that while people make history, they do not make it in circumstances of their own choosing. Only subjects can act, but there is no such thing as a purely individual action. Human action is always an expression of socio-cultural, subjective and material conditions. However, and finally, as the examination of the socio-cultural and material constraints on action has already been undertaken to some extent by the thinkers just named (among others) I see the present task confronting an alternative human geography as precisely one which takes account of action and agency: an action-oriented *social* geography.

These concerns are not only relevant to geographical debates on society and space, but to socio-philosophical and sociological debates, especially those linked to the work of Michel Foucault and Anthony Giddens. For what I am arguing is that we should take Foucault's assertion (1983: 219) that 'power exists only when it is put into action' and that 'power' is exerted only by one to, over or through another. The fact that – for the exercise of 'power' – the body, human corporeality and other material means are of central importance should not lead to the conclusion that 'space', in itself, is something relevant to power, unless 'space' is shorthand for all the material conditions affecting the corporeality of the actor. As I have indicated, these have to be central to any theory of action.

Now after these references to homogenization, difference and the explanation of totalities, the reader may well expect that I am about to launch into an argument for subjective agency based on contemporary post-structuralist sources. But I am not. Instead, I intend to take a different path, back through the social philosophy, social theory and epistemology that has shaped twentieth-century geography, and its definition of what counts as 'science'.

The reason for this is not that I wish to provide a general introduction to epistemology and social theories, especially theories of action (although I shall be glad if the book fulfils that purpose). It is rather that the overt geo-determinism of the nineteenth century has survived in covert forms in both the 'spatialism' and behaviourism of twentieth-century geography. These rely, first, on a definition of 'science' which has to be reconsidered, which means we need to reconsider epistemological definitions of what constitutes a science. I should add here that it is not only the scientistic claims of behaviourist geography that prompt my investigation of epistemology. There is a growing tendency today to think that epistemology and the definition of science are irrelevant to geography, and that epistemological concerns should be replaced with ontological ones. Despite my critique of scientism, I do not agree with this rejection of epistemology: it is difficult to think about ontological categories without first thinking in epistemological ones. Second, the seeds for action theory can be found in thinkers as unlikely as Karl Popper and Talcott Parsons, as well

as more obvious sources such as Max Weber and the phenomenologist Alfred Schutz. In their work we find indications of how
we might develop a social geography, one which places subjective
agency alongside socio-cultural and material constraints. Agency
and action have been in the wings of materialism and functionalism alike, and a close reading of the relevant theorists shows how
often they presuppose subjective agency, and how uncovering
their presuppositions can help theorize it, at the same time this
close reading will show how these theories either implicitly or
explicitly deny subjective agency.

Neither Popper nor Parsons would have wished to see their
work used in the way I shall use it. Yet investigating their logical
implications, and seeing where they take us, has another purpose.
Not only will it contribute to an action-oriented geography, it will
also show that the geo-determinists, behaviourists and struc-
turalists who claim Popper's 'scientific' authority especially are
claiming an authority based on a complete misreading of Popper.
In fact, if I may be permitted one personal aside, I started the
research for this book as a critique of Popper. I was surprised when
I read him that the authority invoked by the kind of geo-
determinism, spatialism and behaviourism I wished to criticize had
apparently not been read by those who invoked him.

In the remainder of this introductory chapter, I shall explore
the antecedents and logic of behaviourist geography. While
'theory' has mattered in geography since the 1960s, the language
of theory, especially with regard to the relation between behaviourism and action theory, has remained largely undifferentiated. Indeed, part of the reason that 'action theory' has
been ignored or dismissed is that it is and has been assimilated to
behaviourism. One of the main aims of this book, in fact, is to show
that the behaviourist and action approaches are opposed, even
though the key concepts of 'behaviour' and 'action' are regularly
interchanged as synonyms. So we should begin by clearly de-
limiting these two terms. Only then can we examine the claim that
research in social geography should be based on action theory and
not on behaviourism. Otherwise such research cannot achieve the
goal it has set itself: to investigate society in relation to its 'spatial'
dimension.

'BEHAVIOUR' AND 'ACTION'

The consequences for methodology and theory of the confusion of 'action' and 'behaviour' can be seen in a comparison between the literature of psychology and sociology since the publication of the two standard works in 1913: J.B. Watson's 'Psychology as the Behaviorist views it' and Weber's 'About some categories of interpretive sociology' ('*Über einige Kategorien verstehender Soziologie*').

If human activity is abstracted according to the definition of '*behaviour*', it is being seen as observable, i.e. empirically perceivable in the true sense of classical behaviourism.[5] All non-observable cognitive aspects of human activity are dismissed, since it is necessary to be able to understand every act in terms of an (observable) 'stimulus' and an (observable) 'response'. Every object in the physical environment represents a potential 'stimulus'. In empirical research, an object is described as a 'stimulus' the moment it affects a behavioural reaction. 'Response' is defined as 'anything the living being does' (Watson 1970: 6). The reduction thereby envisaged of human acts to observable physical processes should, according to Watson and his followers, allow first and foremost a consistent application of (a natural) scientific methodology in the social sciences. The aim of behaviourist research is perforce to define behaviour causally within the framework of scientific theory, such that, given certain 'stimuli', the corresponding response of a living being can be predicted in determinist–nomological fashion.

Theories of cognitive behaviour have been variously applied by 'behavioural geography'. At the same time, they represent a development of classical behaviourism, since behaviour is no longer described solely in terms of stimulus and reaction. 'Stimuli' are conveyed *via* the element of reflection, cognition and consciousness, and are only then considered relevant to behaviour. The cognitive component (motives, needs, attitudes, level of aspiration, etc.) is seen as an interpretational and perceptual filter of 'stimuli'. Stimuli, in turn, are now described as items of information. Within these theoretical terms, human behaviour is explained as responses to stimuli, selectively received from the social and physical environment, which have been cognitively processed into information. In addition, the models of behaviour theory applied in social geography differ from general cognitive theory in that particular emphasis is placed on the spatial dimension. The model for this theory is reproduced in very simplified form in Figure 1.

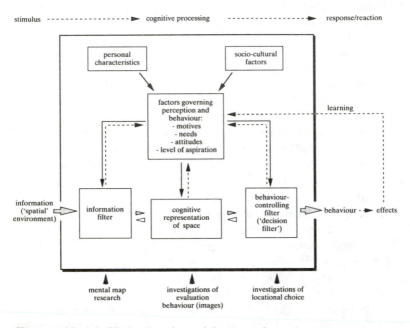

Figure 1 Model of behaviour in social geography

I do not at this point wish to attempt a complex analysis of the behaviourist approach to social geography. Suffice it to say that its basic research foci are evidently compatible with the thinking that makes 'space' into a cause. Bodies react in deterministic ways and are determined unwittingly. Apart from its crudity, an additional problem with this approach is that the results of behaviourist research about 'spatial awareness', 'mental maps' and so on are not related to each other in terms of their basic assumptions concerning individual behaviour (although they could be, if the behaviourists made the effort; cf. Figure 1). The representatives of 'behaviourist geography' sometimes observe that subjective perception of spatial patterns deviate from 'objective' facts. But they go no further. The reasons for different subjective percep-tions, and their effects on various types of behaviour, are not really investigated. Nor is subjective perception related to subjective behaviour. From an 'action' perspective, things look different.

'*Action*' can be broadly defined as a reflexive and intentional activity: a consciously considered, 'freely' performed activity which is goal oriented. This may be brought about by internal (mental) or external activity (observable muscle activity) as opposed to a mere response to stimuli. 'Action' can be defined in its simplest form as 'intentionally effecting or preventing a change in the world' (Wright 1971: 83). 'Act', on the other hand, 'shall designate the outcome of this ongoing process, that is, the accomplished action' (Schutz 1962: 67). If human activity is termed 'action',[6] it is not just the aspect of 'reflexivity', found in cognitive theories of behaviour, which is considered, but also that of intentionality.

I am not suggesting that there are no human activities which lack conscious intention at the time of action. Here one has to determine whether a consciously deliberated action becomes so routinized that it is no longer necessarily consciously planned. If this is the case the activity is described as 'quasi behaviour'; if not, simply as 'behaviour'. But simple behaviour (physiologically and biologically determined reflexes) is of little social relevance.

'Quasi actions', on the other hand, are described by Habermas (1984: 12) as the 'behavioural reaction of an externally or internally stimulated organism, and environmentally induced changes of state in a self-regulated system'. By this Habermas means processes which are described 'as if they were expressions of a subject's capacity for action', where in reality all that is being described is the activity of an organism which itself is not capable of giving reasons for its actions.[7]

Every action (see Figure 2) can be described in terms of four *process sequences*.

1 *Project of the action*,[8] i.e. formation/creation of the intention. This is a preparatory and anticipatory process in a given situation (situation (i)). During this process, the subject is speculating over adequate means for his or her ends,[9] and sometimes over the generalized and justified expectations of other members of a society that have to be fulfilled, etc.

2 *Definition of the situation*: an interpretative sequence in terms of the intended goal. Situation (i) is structured as situation (i'). Here the available means (physical and social) relevant to the goal are determined and selected by the agent. The non-available elements relevant to the intention are the 'constraints'. The situation is interpreted according to certain values

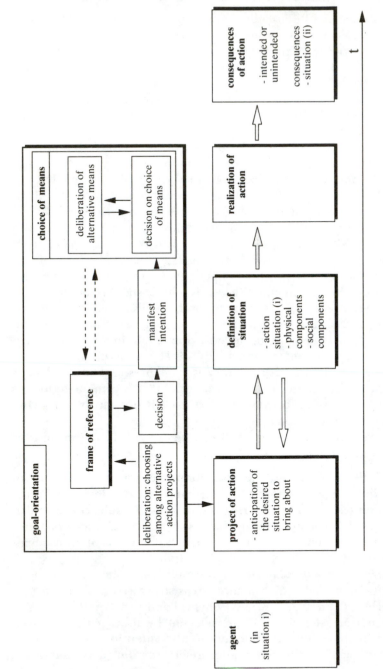

Figure 2 Model of social action

and norms. Sometimes, when the meaning of the situational elements is problematic, a rationalization of their meaning is necessary.

3 *Realization of the action*, or the realization of 'the subjectively imagined goal' (Girndt 1967: 30). This is the implementing sequence of the action, whereby situation (i) is changed or prevented from change. Either the technical component (means–end relation), the legitimacy of the action, or even the meaning component can in this stage be problematic.

4 *Consequences of action*: the intended or unintended consequences of action are constituting situation (ii). This new situation is relevant for the agent and for other agents. For them, situation (ii) might again appear as situation (i). This changed situation can be relevant for forming a 'new' means–end relation, and reinterpreting values and norms as another condition of understanding.

These sequences may not be observable by others (as an intellectual attempt to solve a problem), or they may be 'overt, gearing into the outer world' (Schutz 1962: 67). Behaviour theory explains human activity as determined by stimuli, whereas according to action theory it is intentional and meaningful. Concentrating as it does on the mental processes of individuals, cognitive behaviourism is not able to carry out research into society. This is because it implies that the meaning context of socially relevant activities is reducible to individual stimulus behaviour. It thus cuts out social context. Problem situations appear at best in the light of individual cognitive dissonances. The meaning-context of the social world can only be grasped if we regard the activities of members of society as intentional and not merely as 'responses'. Action theory provides a framework to do this. Behaviourist theory does not.

The basic structure of actions as described above is developed conceptually by Max Weber. He constructed fine conceptual distinctions which cover the manifold forms of institutionalized action. Figure 3 shows these relations in more detail.[10] Within this framework, 'actions' represent the basic units of the process of socialization, and can therefore be regarded as the '*atoms*' of the social universe. They are the smallest unit of investigations of society, the social world. From this it should be obvious that not agents, actors or individuals are the subject matter of action-oriented social research, but actions. Agents are the precondition for actions, but not the research unit.

Another important difference between behaviourist and action theory thematizations of human activity consists in their logical construction. The stimulus–response model implies an inductive logic, whereas the action-theory model implies a deductive structure of knowledge and activity.[11]

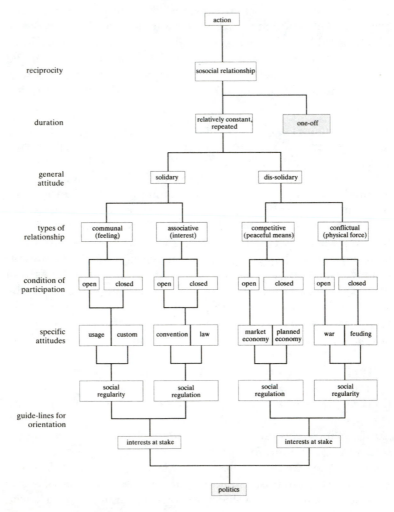

Figure 3 Social relations

BEHAVIOURIST SCIENCE OR ACTION THEORY?

There are two approaches to social geography oriented to human activities. One is oriented to society and action. The other is an individually oriented, behaviourist approach. The one has to be replaced by the other because these approaches are diametrically opposed. Obviously the theoretical concept of 'behaviour' is inadequate for an understanding of social relations and should be replaced. Action theory, adhering to its basic premises (reflexivity and intentionality) cannot logically be reduced to the premises of cognitive behaviourism (conscious interpretation of and response to information).

Yet attempts at precisely this reduction have a long tradition in the social sciences. As I have stressed, they are also to be found in contemporary social research in geography. The reductions start from the assumption that all social phenomena can be reduced to the psychological factors of perception, learning and thinking, with particular stress laid on 'human nature', meaning instincts, instinctive aversions, repressions, anxieties, neuroses, desires, wishes, needs, individual levels of aspiration, motives and the like. Not only Marx, but Popper (1969: 89) opposes this approach with the provocative question: 'Whether it is not the other way around, that is to say, whether the apparent instinct is not rather a product of education, the effect rather than the cause of social rules and institutions?' He goes on to ask whether social phenomena are not 'conventional' in nature rather than 'natural'.

From the behaviourist point of view, a sociologist or social geographer would respond that it is precisely these conventions that are the result of cognitive dissonances between various elements of the motivational structure, or of inadequate adaptation of reactions to stimuli. The principle of such an argument would be: the laws of the phenomena of society 'are and can be nothing else but the laws of [. . .] passions of human beings' (Mill 1973: 877). In other words, social rules must be explained *via* the mental processes of individuals and not vice versa.

From the stand point of the social sciences, this theory is countered by the postulate that no activity can be explained by reference to psychological factors only, but must be explained in relation to social context. Our activities cannot be explained 'without reference to our social surroundings, to social institutions and to their manner of functioning' (Popper 1969: 90). According to the 'sociological approach', it is impossible to reduce social geography to a behaviourist analysis of human activity. Every such analysis must start with the social and not the psychological aspect, because the social cannot be determined by the psychological: it is the social context that is the predominant factor in human actions, and not psychological factors. If any reduction is desirable, it is the reduction of psychological analyses to sociological analyses. Psychological attempts at sociological analysis have to take society as given, whereas action-theory approaches can *explain* the ways societies 'function' and point to difficulties which might impede social actions.

This does not mean, however, that the psychological approach is of no significance for the sociologist. It is just that it should not be seen as the basic concept of social analysis, for the explanation of social phenomena is too much for behaviourism, even the cognitive version. If – as is the case in much social geography – it is claimed that social phenomena can be adequately analysed by means of a behaviourist concept of research, massive problems result. On the one hand, even the developed, cognitive form of behaviourism cannot integrate an intentional, symbolic relation to specific facts into its pattern of 'reaction to stimulus'. On the other hand, every behaviourist argument is forced to *assume* social facts as given, which is incidentally another reason why these arguments are so eager to seize on 'space' as an unproblematic cause. This is also evident in Figure 1: 'socio-cultural factors' are taken for granted. We can conclude from this what by now is more than plain: it is impossible to explain society in behavioural terms. Thus behaviourism cannot become the basis for social research.[12]

Apart from these general arguments against behaviourist social research, I should point out certain inconsistencies and methodological difficulties inherent in it. It is not only in connection with questions of theories of location that behaviourists sometimes speak of observed individual decisions. Since decisions can in principle be taken only with regard to a given objective, behaviourists avail themselves of a manoeuvre which cannot logically be derived from the basic premises of their theory. In the context of behaviourism, reactions resulting from stimuli or information can generally only be described as decisions if the expediency of the observed reaction of an individual 'can be explained as purposeful *activity* [. . .], ascribing this activity to a subject capable of making decisions' (Habermas 1984: 12). If behaviourists wish to explain decisions, they must temporarily abandon their basic theoretical concepts, and avail themselves of action theory (which, of course, they are unlikely to do, unless they are willing to make a paradigm shift on the basis of intellectual honesty rather than academic expediency).

In addition, and this brings me to the crucial crevice between behaviourism and action theory, behaviourists distinguish in effect between the activities and thinking of researchers (those who know) and the subjects they investigate (those who know not). How can one maintain that research activities are intentional, if at the same time the unfortunate subjects under investigation are reacting to stimuli rather than intending to do what they do? Now, of course, one could say that the behaviourist researchers are also reacting to stimuli (promotion, salary), but this does not solve the problem, as these stimuli presuppose a goal-orientation at one and the same time.[13]

These criticisms apply to all the efforts of geographers who seek to base social geography on behaviourist theory.[14] Such efforts derive from the conviction that 'behaviour', as opposed to 'action', is the more comprehensive basic theoretical concept and that it is more adequate to the needs of research in social geography. This conviction need not be explicit. In later chapters, we shall see that many 'action' theories have strongly behaviourist or functionalist leanings. As noted earlier, language here is misleading: the fact that a theory calls itself an action theory does not mean it is consistent in the account it takes of subjective agency, or that it is an action theory at all. For true behaviourists (whether they call themselves by that name or some other) it is not only the conscious, goal-oriented actions that are important in human activity (if they matter at all), but the responses of an individual or group. As 'individuals' are not aware of all their responses, individuals cannot be studied in terms of their goals or purposes. Paradoxically, behaviourism, in this respect, makes an assumption similar to its main opponent in psychology: it assumes, like psychoanalysis, that human beings are governed by forces beyond their control. Yet behaviourism justifies itself, in part, on the grounds that its account of all human activity avoids the 'unconscious'.

In geography, the desire to avoid goals and intentions can be more easily understood if we consider the context in which the rise of 'behavioural geography' took place. The aim of the founders of behaviourist social geography was 'to replace the simplistic and mechanistic conceptions that previously characterized much man–environment theory with new versions that explicitly recognize the complexities of behaviour' (Gold 1980: 3). Alongside the criticism of the traditional geographical presentation of man–environment relations, explanations of human choice of location, based on the 'homo economicus model', were replaced by cognitive behaviourist models. Thus the claims of social geographical behaviourists can be interpreted such that on the one hand human activity should not be described as purposive-rational, when it is not. It 'would be a serious misunderstanding to believe that it is the purpose of model constructs in the social sciences or a criterion for their scientific character that irrational conduct patterns be interpreted as if they were rational' (Schutz 1962: 44). It is far more important to develop empirically useful models of action, so that the objective formulated above can be achieved without falling back on behaviourism.

The cognitive behaviourism of social geography classifies, in its socio-psychological tradition, the reaction of individuals to the physical and social environment in terms of psychological conditioning factors. It is therefore at best appropriate for a geography of the individual. It could not, however, investigate how social phenomena are constructed and produced. The best action theories see human activity as acts through which 'society' is constituted and spatial patterns of material artifacts are produced. Through the analysis of actions, society (and not the individual) can be examined.

But as I stressed earlier, my emphasis on action does not mean I maintain that a given form of society and the spatial pattern of concrete results of action (material artifacts), produced by its members, are wholly derived from the conscious objectives of individual agents. The dimensions of 'reflexivity' and 'intentionality' contained in the concept 'action' should not be misunderstood as meaning that all the *consequences of action* are *intended.* The action-theory approach takes as its starting point the fact that the social world, the agent's social context, is the result of human actions and decisions, and that it can be changed by subjective agency alone. This does not mean, however, that any current form of society and the current shape of the spatial pattern of towns, transport routes, regional disparities between infrastructures, etc., are the result of a conscious human plan. In the first place, plans of action are not always realized in the way they were envisaged. Further, the consequences of an action, even if the plan was realized precisely as envisaged, are not always those predicted by the plan. In addition, even in cases where plan, realization and consequences relate to each other precisely as envisaged, the consequences can affect agents unfavourably, which again was not their intention. Finally, these consequences can impose constraints upon others, so that they can no longer realize their own objectives.

The idea of action-theory research into social phenomena should not therefore be misinterpreted as a 'conspiracy theory of society' whereby all social problems are consciously caused by a certain group of agents and consciously maintained for their own benefit. It should be understood rather as an alternative to the behaviourist approach, which can only investigate social phenomena by means of inadequate reduction processes, either taking for granted or ignoring the actual object of research: the social world.

An action-theory analysis of society sees human actions, but not the subjects who perpetrate them, as the basic category of social analysis. Society is understood as the totality of actions in their orientations, realizations and consequences specific to different cultures, institutions, groups, etc. *The general aim of action-theory research is to classify the complexity of social phenomena and problem situations in the light of the actions of the members of society.* More precisely: to understand, explain and, where there are problem situations, put forward suitable suggestions for bringing about change in the kind of actions causing the situation. The power at the disposal of different agents of course varies considerably, their actions and methods of action producing a greater or lesser social effect. Whether intended or unintended, the consequences of former actions must always be understood as the conditions, constraints and/or means for the contemporary human agent.

For example, if in this context we regard the spatial pattern of infrastructural institutions as the result of human actions, it is obvious that we need to know the orientation and meaning of actions and their course. Only then can we understand their present tangible effects and, if required, make appropriate changes to the relevant institutions. If we wish to understand the significance for the social world of geographical areas which have been shaped or protected by human hands, we must first be in a position to give an adequate explanation for human activity in its social context. If such activity is more powerfully determined by social factors than by those of space and distance, we shall learn more about the social significance of 'spatial' factors by first analysing the social factors.

If we accept these arguments, the question remains open as to what criteria should be met by an action-oriented theory of social geography. Apart from contributing to the successful solution of problems of everyday practice, such a social geography also has to meet some methodological criteria for empirical research. I do not reject the necessity for this, although as I indicated, the attempt to find acceptable research criteria has been more or less by-passed in the theoretical critiques with which I have most sympathy. In fact, trying to establish workable criteria for research is something that is constantly in the forefront of this book. Moreover, the following chapters will show that the contemporary debates about the return of the subject and the place to be given subjective agency are not putting new concerns on the theoretical agenda; they are rediscovering concerns that have figured throughout epistemology and social theory for a century. It is with this marginalized aspect to mainstream epistemology and social theory that I shall be preoccupied. This means reading sources that are often ignored today for the insights they shed on these 'contemporary' theoretical issues.

Taking this from another standpoint: the social scientists who believe that social science should be modelled on the natural sciences derive their criteria from that most esteemed epistemological source, Karl Popper (they think). Partly, but not only, for this reason I shall discuss Popper's epistemology in some detail. An investigation of it shows that Popper's 'objective' criteria for the natural sciences differ from his criteria for the social sciences. I draw this out in Chapter 2, in a discussion of what I have termed the 'objective' perspective. This discussion is also useful because it shows that the three components of action (social, subjective and physical), already mentioned, are mirrored in Popper's 'three-world' theory of knowledge. Thus Popper, whatever else one might think of him, is relevant to developing an action theory, especially in a discipline like geography, which takes account of both the social and natural worlds.

Of course Popper is not the only epistemological antecedent geographers and social scientists call on. The main tradition opposing the 'objective perspective' comes from phenomenology. In Chapter 3 I turn to this 'subjective' epistemology, especially to the work of its leading figure in social thought, Alfred Schutz. Here again, we find three (subjective, social and physical) factors figuring in epistemology, and the contradictions between the objective and subjective perspectives are not as striking as they appear to be at first. In fact, as Chapter 4 will show, these perspectives are often complementary.

After discussing epistemology, I turn, in the remaining chapters, to the sociological action theories of Pareto, Weber, Parsons and Schutz. I also discuss the rational choice theories devolving from one tradition in economics. This discussion has two aims: one is showing that 'action' theories often leave out agency; the other is culling these theories for what they can contribute to a theory of action in geography, a theory which ties back to the social, subjective and physical criteria, and opens out new research directions. Crucial here is the fact that the agent, actor or subject is always embodied, and this corporeality affects whatever position he or she might adopt. But it does so not because the subject is nothing more than a physical being; rather, actors can move from one point to another, depending not only on what enables them to move (the constraints or facilitating factors in agency) but on what it is that they want. From there, we return to the space question, and the discussions which reduced social and subjective factors to physical ones in the first instance.

Chapter 2

The objective perspective

In his critical analysis of 'common sense' theories[1] Popper states (1979: 1ff.) that they have previously been characterized by two features: the tendency to acquire knowledge according to the principles of the 'bucket theory of the mind' (often appealed to when reference is made to 'common sense'), and the underlying idea that there exists a real world independent of the knowing subject. The first feature is considered by Popper to be the 'mistaken part of the commonsense theory of knowledge' (1979: 39), which he cannot accept. The realism postulate (which we shall see he relocates), the assumption of a real world, he considers a necessary presupposition for research, even though it can be neither proved nor disproved. Popper's critical rationalism is often referred to as the 'realistic' or 'objective' perspective. He starts from the premise that both the physical and the social world exist independently of the agent's mental world. According to this view, both worlds consist of objective phenomena. So the basic postulate is: social phenomena have the same objective existence as physical phenomena.

POPPER'S VIEW OF TRADITIONAL COMMONSENSE

When Popper describes common sense as the 'bucket theory of the mind', he means this: all representations of commonsense acquisition of knowledge have assumed that this acquisition corresponds to an inductive theory of knowledge.[2] He draws on examples from the positivist tradition of knowledge from Hume onwards. Hence, 'there is nothing in our intellect (or "bucket") which has not entered it through the senses' (Popper 1979: 3); the 'bucket' is (consequently) empty at birth; subjective perceptions form the precondition of all consciousness and all knowledge; the most certain knowledge is the knowledge that enters the 'bucket' directly, i.e. without the aid of the perceiver (e.g. his or her preconceived ideas, prejudices, etc).

This inductive interpretation implies that knowledge is the 'relation between the subjective mind and the known object' (Popper 1979: 146). According to Russell, by means of habit, through repetition and the mechanism of association of ideas – generalization – a 'belief' is derived from knowledge, belief in certain rules or laws in life. The more directly, frequently, and regularly a knowing subject has the relevant experiences, the more expectations which result from it become fixed and certain. In this traditional commonsense theory of knowledge, which Popper criticizes, certainty of, and justification for, the expectation of regularity depend therefore on the frequency, repetition and directness of the experience. Sensory impressions are seen as the most certain basis for knowledge. The aim of knowledge, according to the positivist proponents of this approach, is the achievement of the absolute certainty that expected events and relations will actually come to pass.

POPPER'S CRITIQUE OF THE INDUCTIVE INTERPRETATION OF COMMONSENSE ACTIONS

Popper, in his critique of positivism, distinguishes between two aspects of Hume's (1978) formulation of the induction theory. Popper's first criticism is that it is not logically justifiable to infer *probable* future instances from repeated instances already in our stock of experience or 'bucket', however great the number stored. Hume suggested that the inductive process should be regarded as a correct description of common sense, in spite of its logical flaws: for him, common sense accordingly is an irrational theory of knowledge. Hume thought that the decisive argument was that a human being would be unlikely to survive without the beliefs derived from repetition and the association of ideas. According to Hume, 'our "knowledge" is unmasked as being not only of the nature of belief, but of rationally indefensible belief – of an irrational faith' (Popper 1979: 5).

Popper does not accept Hume's conclusion because he is not prepared to accept that human understanding operates irrationally. Popper does not deny that we act on the basis of pragmatic convictions, but he does not allow that these are 'the irrational results of repetition' (1979: 27). In order to argue that the process of knowledge is rational, Popper starts out from what he terms the 'transference principle'. This presupposes that anything valid in logic must also be valid in psychology; that is, human understanding can only attain knowledge in a logically valid, i.e. rational, way. This is the key postulate of rationalism. If this is accepted, it is no longer necessary to assume, as Hume did, that an irrational belief based on the associations of sensory impressions is the basic principle for action. On the contrary, knowledge can be seen as the result of logically valid deductions from given pragmatic beliefs, i.e. dispositions and expectations.

In contrast to perceptions, pragmatic beliefs are 'partly *inborn*, partly modifications of inborn beliefs resulting from the method of trial and error-elimination' (Popper 1979: 27) 'through learning' (Popper 1979: 275), and are therefore not the result of amassed sensory perceptions. Popper also refers to these beliefs as knowledge or a system of dispositions, which are however never entirely certain but always conjectural or hypothetical in nature (Popper 1968: 98–9). The basic thesis is: 'All acquired knowledge, all learning, consists of the modification (possibly the rejection) of some form of knowledge, or disposition, which was there previously, and in the last instance of inborn dispositions' (Popper 1979: 71). Since at least parts of this knowledge precede any apparent perception (e.g. breathing, swallowing, movement of limbs, etc), they act as a kind of selection principle for the perceiving or experiencing subject. Thus, for Popper (1979: 24), every perception or experience is preceded by an interest or a problem. Accordingly, every perception or experience already embodies a proposition, and to this extent, a theory.[3] What is regarded by the subject as highly important input, and what is ignored, always depends entirely on the dispositions present, i.e. the theory. All our observations may thus be seen as 'theory-impregnated' (Popper 1979: 72). The process of understanding, human thought and the *action* resulting from it, should therefore be regarded as rational, i.e. logically valid. The starting point and basic requirement for knowledge is always a given background knowledge in the form of *hypothetical* expectations. In contrast to inductivist, positivist ideas, logically valid deductions are based on actions adapted to real, objective surroundings.

The arguments of the inductive, positivist theory of knowledge are therefore in Popper's view flawed at the very outset. For they imply that there can be nothing that is purely experience or purely perception. Popper thus calls his description of the process of human knowledge the 'searchlight theory' of knowledge (Popper 1979: 341ff.). What can be illuminated by our knowledge at any one time by a hypothesis, i.e. by our theory, is observed and may be understood. The rest of reality, which exists independently of us, remains in obscurity. All theories should be regarded as 'dispositions to act, and thus a kind of tentative *adaptation to reality*' (1979: 41), to our surroundings. If these theories are correct, i.e. if they coincide with reality,[4] we can (for the time being) survive. If they are incorrect they must be abandoned, unless we wish to perish like the creatures who became extinct during the course of evolution, owing to their inability to adapt. The second basic proposition is: 'All growth of knowledge consists in the improvement of existing knowledge which is changed in the hope of approaching nearer to the truth' (1979: 71). Thus it is not only the process of knowledge and human thought that should be regarded as rational and logically valid – not irrational, as Hume believed – but so should human action. For 'we do act not upon repetition or "habit", but upon the best tested of our theories which [. . .] are the ones for which we have good rational reasons [. . .] for believing them to be the best approximation to the truth' (Popper 1979: 95).

RESEARCH FROM THE CRITICAL RATIONALIST STANDPOINT: GENERAL REQUIREMENTS

We have now outlined the basics of research from the critical rationalist standpoints. Popper (1979: 72) accepts knowledge resulting from everyday observations only as a provisional starting point for scientific activity – as 'background knowledge' or a 'horizon of expectation'. This point has to be elucidated, however, and to this end criticism is decisive. Popper wishes therefore to be regarded as 'a critical commonsense realist' (1979: 323). He argues that any scientifically acquired knowledge, in natural and social science alike, should be understood as an elucidation of common sense.[5] From the perspective of the objective theory of knowledge, it is the task of science to elucidate critically, to bring our knowledge closer to the truth. If we start from the premise that all our subjective knowledge is theory-impregnated, it becomes obvious that it can never be 'pure' in the sense of 'certain'. It is also clear that certainty cannot be achieved by a further amassing of data. Truth, or closeness to truth, cannot thus be attained by an increase in empirical data. An increase in data should not strengthen belief in expectations, as the positivists claim. Truth is rather achieved by a process, a method which adopts a critical attitude, and which is committed to critical discussion. The elimination of error in the sphere of everyday life is complemented in the scientific sphere by conscious *criticism*, with its controlling idea of the search for truth: 'Criticism consists in the search for contradictions and in their elimination' (Popper 1979: 126).

The critical method can be more precisely defined with the aid of two distinctions between positivist empiricism and classical rationalism. First, there is the recognition that for the reasons given above we can never achieve sure and certain knowledge of external facts. The interpretation of knowledge offered by naive and logical empiricism, i.e. neo-positivism, is thereby invalidated. Second, classical rationalism founders on the fact that, according to Popper, no axiom from which deductions might be derived can be based on evidence. In addition, no deduction increases the information content of the propositions. This process 'merely' preserves the truth. Popper suggests as a solution a deductive method of testing empirical theories. The stricter the tests, and the longer the hypothetical propositions withstand refutation, the closer they will be to the truth. For according to Popper there is no direct access to truth, only an approach to the truth *via* the elimination principle of falsification. But in order to establish whether a (hypothetical) proposition is refuted or not, a statement has to be capable of being disproved by a test. It has to be falsifiable. Thus Popper's famous criticisms of psychoanalysis and Marxism rest on their ostensible non-falsifiable claims. An explanation of this falsification test is complicated, however, by the fact that Popper (1979: 323) maintains that in contrast to the empiricists he is 'a realist in two senses of the word'. What does he mean by this?

This brings us to the vital importance of Popper's realism postulate. This postulate is linked to the idea of an objective reality independent of all knowing and acting subjects. He sees this as the idea which regulates all scientific activity. On this basis statements can be presumed to be (provisionally) true, or finally rejected as false.

In more detail, Popper's objective theory of knowledge is based on the distinction between three 'worlds'. These worlds have different ontological status.

> The first [world 1] is the physical world or the world of physical states [and objects]; the second [world 2] is the mental world or the world of mental states [world of states of consciousness, dispositions to act]; and the third [world 3] is the world of intelligibles, or of *ideas in the objective sense*, it is the world of possible objects of thought: the world of theories in themselves, and their logical relations; of arguments in themselves; and of problem situations in themselves.
>
> (Popper 1979: 154)

It is the fact that world 3 is available to them which, in Popper's view, distinguishes human beings from the rest of nature.[6]

World 1 and world 3 are independent of the subjective world 2; that is, worlds 1 and 3 are objectively given. But they are only linked to each other *via* world 2. Thus the world of agents' subjective states of consciousness assumes a mediatory function in so far as the first and the third do not affect each other except through the world of agents' personal experiences. World 3 is indeed 'man-made and "super human" at the same time. It transcends its makers' (Popper 1979: 159). It is independent of them, just as spiders' webs (as products of spiders), once they are finished, exist independently of spiders.[7] It is the same with a scientific theory or its discussion. As long as the hypotheses and/or arguments remain in the consciousness of the scientist, as long as they are only thought, they belong to world 2. But as soon as they are communicated they take on an objective character (world 3), and from this moment on they are no longer dependent on the subject, but have logical content of their own. Frege (1892: 32) formulated this independence as follows: 'I understand by a *thought* not the subjective act of thinking but its *objective content*.'[8] World 3 comprises the whole of (objective) knowledge at a given time. It also comprises the knowledge and possibilities that are as yet unknown.

'Human language, and thus larger parts of the third world, are *the unplanned product of human actions*' (Popper 1979: 159), so that new problems may always arise as unintended by-products of those actions. We do and can constantly discover new problems, new hypotheses, and new critical arguments in world 3 which were there 'before anything corresponding to them appeared in world

2 [. . .], in a similar way to that in which we (*sic*) can make geographical discoveries in world 1' (Popper 1979: 74). To take another example, the problem of primary numbers was discovered after the invention of numbers. In order to solve such problems we may invent new theories as products of our critical thought, and these in turn produce new, unintended and unexpected independent problems which have to be discovered, etc. Popper's main idea here seems to be that the state of knowledge (world 3) should be seen as a given state of closeness to *objective truth* in itself. This state of knowledge is the result of the critical elimination process of meaningful hypotheses effected through falsification. It is the task of scientists to extend this process if there is to be any progress in scientific knowledge.[9]

For the elimination process, the empirical verification of hypotheses through observation and the critical arguments of world 3 comprise the 'tests' by which statements can be refuted. The boldness of a hypothesis, and thus its falsification potential, is the greater the more information it contains. Popper sees hypotheses as thematically and logically systematized sets of (provisionally) true nomological if/then statements with information content.[10] The information content increases as the hypothesis' 'if' components become more general, and its 'then' components more specific and precise. The hypotheses formulated must however also be meaningful in relation to current knowledge (world 3), if growth of knowledge is to be achieved. Crucially, Popper's postulate of an 'objective epistemology without a knowing subject' means that the growth of knowledge is no longer dependent on the conscious subject, but on world 3: the advance towards an absolute truth independent of any subject. Moreover, Popper sees possibilities for the practical application of the results of scientific action in the sphere of everyday actions. These consist in the fact that (provisionally) valid precepts for action can be derived from the scientific theories, making possible a more successful *intervention* in physical and social reality.

To sum up: for Popper the general, all-embracing goal of science is constantly to test existing theories. In this way scientific knowledge will move closer to objective truth, and he proposes criticism, i.e. the deductive method of falsification, as the means of achieving this goal. In this Popper breaks with all notions of 'certainty'. His conception of science can be understood as 'rationalism'. *Contra* Hume, Popper's conception of knowledge is that it is logically valid, and rational. Rational procedures are, however, not only accessible to *a priori* criticism, i.e. axioms, but also to empirical criticism: the criticism of reality, on which hypotheses can founder. We must therefore call it 'critical rationalism', that is, rationalism which can be empirically revised and tested. Such revision and testing lies in finding observations which contradict the hypotheses, or by means of arguments from world 3 brought forward by other scientists. On the other hand, it is seen in criticism by means of observation statements from the sphere of the physical and social world, which are used to test these same hypothetical assumptions.

FUNDAMENTAL PRINCIPLES OF CRITICAL RATIONALIST METHOD

Popper's view presupposes the existence of a universal 'eternal' truth which is independent of the agent. Scientific research should approach this truth as closely as possible. The procedures used, i.e. the scientific method of critical rationalism, can be described as 'the method of bold conjectures and ingenious and severe attempts to refute them' (Popper 1979: 81). The constituent postulates of this view of scientific method can be summarized as follows:

- *Realism postulate.* There is a real world which exists independently of the agent; it is the final test of the truth of our hypotheses.

- *Transference postulate.* What holds in logic also holds in psychology and therefore in human action.

- *Biology postulate.* The vast majority (99.9%) of our theories are inborn, i.e. inherited, and only 0.1% 'consists of the modifications of this inborn knowledge [. . .] and *the plasticity needed for these modifications is also inborn*' (Popper 1979: 71).

- *Rationalism postulate.* Human knowledge and action are based on the principle of deduction, and not of induction; processes of knowledge and human action are therefore valid in the logical sense: that is, valid in formal, but not always in empirical, respects. They are rational and not irrational.

- *Criticism postulate.* Our knowledge, from which we (deductively) derive the conclusions which affect our actions, should be

regarded as provisional and hypothetical, and always be subjected to further criticism.

For Popper (1979: 71) the significance of objective knowledge for
world 2, the world of the agent's consciousness, is most evident in
the fact that world 3's action upon us has become a more powerful
evolutionary force than our creative action upon it. World 3's
effect on world 1 is also considerable, since it is transmitted by
subjective action (world 2): 'One need only think of the impact of
electrical power transmission or the impact of atomic theory on
our inorganic and organic environment, or of the impact of economic theories on the decision whether to build a boat or an
aeroplane' (Popper 1979: 159). World 3 is thus for Popper of
immense importance to the agent. Commonsense actions are also
connected with world 3 insofar as they rest on scientific pre-
sumptions. The latter are, however, far more consciously and critically orientated towards world 3 ('enlightened common sense').
Many world 3 theories, if their formulation is too difficult, will not
enter everyday life, but they are nevertheless no less true or false
than other commonly accepted world 3 theories: all knowledge is
hypothetical, and all observations and actions are directed by
hypotheses or theories. 'All work in science is work directed
towards the growth of objective knowledge' (Popper 1979: 121).

ACTION AND THE ONTOLOGICAL DISTINCTION
BETWEEN THE WORLDS

The interrelation of worlds for the agent can be shown graphically
(Figure 4). The model takes special account of the basic assumptions mentioned: Popper's 'theory of objective knowledge without
a knowing subject'.[11] It also highlights those aspects of his theory
which, I shall argue, bear on action.

In the *action situation* these three worlds,[12] despite their
different ontological status, are plainly related: world 3 and world
2 (a/d) and worlds 2 and 1 (b/c) affect each other. The world of
states of consciousness (world 2) has a reciprocal relationship with
the two other worlds. Thus worlds 1 and 3 can only influence each
other through the mediation of world 2, i.e. through acts of
understanding and of practical action affecting the external world.
In this view, world 3 is not merely an expression of world 2, as
idealist theories maintain, and world 2 is not merely a reflection of
world 1, as asserted by naïve logical positivism.

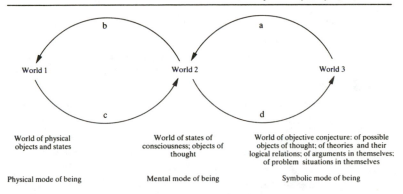

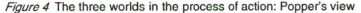

World 1	World 2	World 3
World of physical objects and states	World of states of consciousness; objects of thought	World of objective conjecture: of possible objects of thought; of theories and their logical relations; of arguments in themselves; of problem situations in themselves
Physical mode of being	Mental mode of being	Symbolic mode of being

Figure 4 The three worlds in the process of action: Popper's view

But exactly how does a human being act in an everyday situation? Popper does not really give a direct answer to this question in terms of his three-world theory. However, that theory can be reconstructed in such a way that it can be applied to a concrete situation. In turn, this application leads directly to Popper's concept of society. In 1962, Popper distinguished a physical and a social aspect to a 'situation', in accordance with his three-world theory. The physical aspect, the physical world, 'contains physical resources which are at our disposal and about which we know something, and physical barriers about which we also know something (often not very much)' (Popper 1976: 103). The social components are 'other people, about whose goals we know something (often not very much), and, furthermore, *social institutions.* [These] social institutions determine the peculiarly social character of our social environment' (Popper 1976: 103). In addition, an action situation will bring the agent's mental or subjective consciousness into play (world 2). Since the mental, subjective world is fundamentally directed by components of world 3, Popper subsumes them (1979: 165) without further distinction under 'the problem situation of action': subjective factors are regarded as a part of the objective situation. In a certain everyday situation, an agent can

> for the purposes of social sciences be viewed as in pursuit of certain goals or aims, within the framework of natural, social, psychological and ethical circumstances. These circumstances constitute both the means of achieving [the agent's] aims and constraints on that achievement.
>
> (Jarvie 1972: 4)

In so doing the agent is acting according to the rationalism postulate. 'When faced with the *need to act*, on one theory or another, the rational choice was to act on that theory – if there was one – which so far had stood up to criticism better than its competitors had' (Popper 1986: 103–4).

We must now more precisely distinguish between two meanings of the word 'rational'. The first characterizes the procedures of knowledge acquisition and actions as logical. Agents start out from a theoretical background, an idea or belief, and proceed to a performance of the act, following those theories which they consider to be the best tested. 'Rational' here is not meant in what can be termed an 'empirical sense'. It is not meant in the sense that in a given situation, an agent always chooses the optimum alternative and does the right thing. In fact Popper thinks such cases are very rare (1967: 149), for the theories available to the agent are always incomplete and uncertain. Popper also maintains that human agents frequently act in a rational way in that they learn from their errors. Errors are refutations. They are actions which have failed. This presupposes that human beings often have a rational attitude, even if their actions are not empirically appropriate. In this sense, 'rationality, as a personal attitude, lies in the disposition to correct our ideas. In its intellectually most developed form, it is the disposition to examine our ideas in a critical spirit, and to revise them in the light of critical discussion with others' (Popper 1967: 149, my translation).[13]

The *postulate* of rationality is thus not necessarily or even primarily concerned with the empirical correctness of an intervention in the physical and social environment. It does not refer to the empirical validity of the premises of the agent's theory, but to the claim that agents act according to the deductive principle (of falsification). Agents start out from the more general knowledge that they hold to be true, and from the conditions of the situation, and so select from the means available those which they consider most likely to achieve their goals. If the action fails, agents will, in line with the rationality principle, reformulate the theory or more general knowledge which they formerly held to be true. If the action succeeds, their theory and general knowledge will, for the time being, be retained unaltered.

The *physical–material* aspect of a situation is perceived, related to themselves, and experienced by agents in relation to their expectations, hypotheses and interests. Since every action takes place within the framework of particular material circumstances, this experience is always limited by an objective component which no successful action can evade. Even if this limitation does not play the same role for every action, it still remains an important condition of acting: as well as the physical environment. The body of the agent must also be taken into account. In general, the physical–material limitation leads to a paucity of available means for achieving a goal. From this Popper concludes that *every action is an attempt to solve or to prevent a problem.*

The most important aid to agents in their attempts to overcome physical–material constraints is the technology known or available to them. As a means–end relation, it depends both on the available theory, expectation or idea and on the elements of the situation as they are understood and experienced. Popper believes that there is always a given interrelation between the goals/decisions of the agent (world 3/world 2) and the facts of the physical–material situation (world 1) of the agent. It follows from this that the physical–material elements of the situation alone cannot determine action (which geo-determinists and other extreme materialists assume they do). Nor can the goals of actions be realized independently of this interrelation.

The *social aspect* of the situation is neither physical nor psychological in nature, nor is it an amalgam of the two. For Popper the social world forms an important component of world 3, together with arguments and scientific theories: it is the world of social institutions. Like the other components of world 3, it should be seen primarily as 'the unplanned product of human actions', 'as the unintended consequence of intentional actions' (Popper 1979: 159), where the unintended by-products (even institutions, traditions) exist independently of the agent. A good argument for this independence postulate, in the context of the 'objective perspective', lies in the fact that social institutions constantly manifest unintended consequences, creating new, unexpected and autonomous problems which often become important constraints on human action. The impact of world 3 – and thus also of the social world – on the agent is made clear by Popper (1969: 89) in his agreement with Marx's statement, in the 'Preface to A Contribution to the Critique of Political Economy', that it is not the consciousness of man (world 2) 'which determines his existence, rather, it is social existence that determines his consciousness'. Popper is here emphasizing that agents make decisions on the basis of their world 3 ideas, theories and beliefs, which are institutional in character. Or, in Marx's words, 'men make history, but they do not make it in circumstances of their own choosing' (Marx 1960: 115). Hence institutions and traditions are retained or changed in a partly conscious, partly unconscious way.

For Popper (1969: 90), the 'action situation', as I term it, is thus mainly determined by social, world 3 factors. The question then becomes: How is it that agents are able to harmonize their different means and ends with one another? Popper recognizes that this coordination is possible in principle; every agent orients himself or herself more or less explicitly towards the components of world 3, and towards its social elements in particular. Also, human actions are the connecting links between and within institutions (world 3). Institutions (e.g. markets, unions, science, racial hatred, money, . . .) and the ideas associated with them (world 3) produce states of consciousness and goals (world 2) towards which agents orient themselves. Thus institutions function as conventional means of action coordination, which at the same time constrain our actions. They operate as 'channelling' frames of reference for the orientation of human action, whether its intervention is in the physical or the social world.

If human beings are not to be foiled in their actions by the social world, they must be able to cope with it in such a way that they can achieve what they want and prevent what they do not want. 'We might say that to do this they construct in their minds a conceptual map of the society and its features, of their own location among them, of the possible paths which will lead them to their goals, and of the hazards along each path' (Jarvie 1972: 161). Although the different maps do not tally completely, they make adequate coordination possible as these maps are made up of elements from world 3. They are therefore objective. Those parts of the *social* 'conceptual map' with objective contents which deviate from the others can be constantly tested for truth or falsity, in a continuous intersubjective process (cf. Bourdieu 1985). Since the facts of the social world are constantly changing, it is necessary to run a continuous check on one's own 'map'. Elements which have proved to be false or untrue, those which inspired unsuccessful actions, must be eliminated. The other elements which inspired successful actions should be provisionally retained. Of course no agent ever has a perfect 'map'; this is a science fiction phantasy, underpinning the more manic dreams of artificial intelligence planners.

Actions can thus be foiled or constrained by both the material world and the social world. Those which relate to the social components of the situation should be understood as problem-solving exercises in the same way as those directed towards the physical world. They are attempts to achieve a goal against the background of a particular theory, even if that theory is empirically sketchy. The incompleteness of the theory, the conceptual map can lead to an agent's false or faulty interpretation of the actual social situation. Actions should therefore be seen as trying out possibilities for achieving a goal.

But even if the goal is achieved, the agent often has to continue to try out further possibilities. For it may turn out that the goal achieved does not correspond to what the agent had in mind, or that it is merely a sub-goal, part of a larger goal. By 'trying out', agents clarify their own aims. They acquire knowledge about their social environments, and by this gain experience regarding the actions which can be carried out in a society. Initial partial knowledge and any false interpretations of the situation which result from it are thereby modified and enlarged.

The difference between common sense and scientific action is, as we have already seen, that the latter is more 'enlightened'. Therefore the agent is in a position to judge whether the common-sense action is correct or incorrect. The agent has chosen, as a criterion of judgement, the appropriate means to a given end in relation to the desired goal. Scientists or researchers deduce the most appropriate means to an end from theories which have proved successful hitherto. They then recommend to other agents that they include these means in their plan of action, when they are trying out various means to achieve their desired goals.

I shall now deal in more detail with the question of how, according to Popper, the various aspects of the action situation should be approached.

THE SCIENTIFIC APPROACHES TO THE PHYSICAL AND SOCIAL WORLD

Bearing in mind Popper's theory of the three worlds, we can now examine his remarks on the empirical sciences from the standpoints of developing an action theory. In the next section we shall see that the natural sciences deal with facts and problems of the physical world, with facts of the action situation which lack intrinsic meaning. The empirically valid theories of the natural sciences shape the directions of those actions which intervene in the physical world. On the other hand, of course, the social sciences carry out empirical research into the facts and problem situations of the social world. The results lead to directions for those actions which relate to the social world.

For Popper, the aims of all empirical research are, first, to put 'forward statements, or systems of statements and test them step by step. In the field of the empirical sciences, more particularly, [the scientist] constructs hypothesis, or systems of theories, and tests them against experience by observation and experiment' (1968: 27). The specific task of empirical research can thus be described as the testing of hypothetical statements by experience. Second, the aim of empirical science should be 'to find *satisfactory explanations*, of whatever strikes us as being in need of explanation' (Popper 1979: 191). In the most general sense such explanations should be produced, and the testing of theoretical knowledge carried out, through the application of the falsification principle.

That much said: contra widespread opinion (shared by the proponents of the geographical 'spatial approach'), I suggest that Popper is proposing different objectives and methodologies for the natural and social sciences. Scientific procedures should accordingly be adapted to the respective *ontological status* of the physical and social worlds. This distinction is especially pertinent to ecological research.

In the following analyses of the critical–rational principles for action research, I shall seek to identify and characterize the features of the different procedures used in analysing the physical and social worlds. What they have in common is the method of falsification. This, as indicated, should be used in Popper's model of action: 'trying out' theories and ideas as attempts to problem solving. *Just as actions are based on the basic principle of deduction, so they can be explained according to the deductive method of inquiry.* My thesis is that Popper's postulate of the 'unity of method' (1960: 130), 'that is to say, the view that all theoretical and generalizing sciences (should) use the same method, whether they are natural sciences or social sciences', only relates to the application of the method of falsification. It does *not* relate – as maintained so far – to causal explanation.

Empirical research into the physical world

The theoretical aim of the sciences of the physical world should be pursued without regard to value judgements. Hypotheses should not be prescriptive in nature, and research should be inspired by nothing but the desire to approach nearer to objective truth 'in itself'. The main aim of research into the physical world[14] is, as noted, to make 'ingenious and severe attempts' to refute scientific theories about world 1 which have thus far been held to be true, and which have provided a framework for bold and ingenious conjectures (context of discovery). In this way we approach nearer to objective truth, and *basic research* can be directed by theoretical interests alone. The relation of world 3 to world 1 should be investigated by means of causal explanation, in such a way that the explanations put forward can be empirically refuted (context of justification). Accordingly, the second main aim of the empirical natural sciences is to provide causal explanations of problems in the physical world.

In order to draw out the difference between Popper's arguments about the universal research relevance of falsification, and the ostensibly universal claims about scientific method, it will be necessary to review briefly certain key Popperian concepts: namely, (i) explanatory statements, (ii) universal law and (iii) the concept of 'primary condition'.

The scientific causal explanation represents one form of investigative deductive procedure. A logically valid causal explanation always involves the derivation of a problematical state of affairs or event from a true law (the nomological hypothesis) and the conditions of application of that law. By a causal explanation 'is meant a set of statements by which one describes the state of affairs to be explained (*the explicandum*) while the others, the explanatory statements, form the "explanation" in the narrower sense of the word (the *explicans* or *explanans* of the *explicandum*)' (Popper 1979: 191). It thus comprises two types of statement:

- the *explanandum*, a statement to be explained, which usually consists of a true statement about a (problematical) empirical isolated fact; and
- the *explanans*,[15] which comprises the explanatory statements and in turn has two components: a proven nomological hypothesis, i.e. a law which has not yet been falsified, and a singular statement which renders the law applicable, the 'initial

condition', as Popper (1979: 350) calls it, or the 'primary condition', as it is generally termed.

According to Popper (1979: 191), the *explanans*, in contrast to the *explanandum*, is usually unknown: 'it will have to be discovered. Thus, scientific explanation, whenever it is a discovery, will be *the explanation of the known by the unknown'*, where the *explanandum* follows from the *explanans*.

The universal law. Laws are those statements which have the form 'if . . . then' or 'the more . . . the more', asserting in this way something that has been verified at least once and has since undergone several falsification tests (which it passed without exception). They must not be limited either in space or time, i.e. they must claim universal validity, and must be falsifiable by experience, i.e. reality. In other words, they 'must be *independently* testable' (Popper 1979: 192). If they are to be empirically testable, they must also have information content. That is, they must be statements about reality.

The primary condition. Before the universal law can be applied (to the explanandum), it must first be established whether the condition required by the 'if' component of the hypothesis is present or not in the case in question.

A further aim of science is to find better and more testable empirical theories. This demand has two implications. *First,* the greater its degree of universality, the easier a hypothesis is to test. As the degree of universality increases, so the range of possibilities of falsification increases. *Second,* a hypothesis is more easily testable, the greater the degree of its precision, its information content. For the more precise the statement, the more information it contains. According to Popper, a hypothesis contains information if it makes an assertion about a state of affairs existing in 'reality'. The information content increases according to the number of occurrences which the hypothesis stipulates may not take place. A further aspect of testability arises through the replacement of qualitative statements by quantitative ones, whereby *measurement* becomes an important aid in increasing the testability of statements. The device of measurement becomes more and more important in the course of scientific progress. But 'we must [. . .] not overlook the fact that all measurement is depending on theoretical assumptions. [And measurement] should not be used (as it often is) as a characterizing feature of science or of the formation of theories in general' (Popper 1979: 356–7).

Scientific theories as a frame of reference for action

If they are to succeed, actions which intervene in the physical world must be based on empirical knowledge of natural facts. That is: successful actions must be deductions from empirically valid scientific theories. In their objectives and in the choice of means they must be so devised and carried out that they are adequately adapted to the physical conditions of the situation.

The *practical task* of science consists (Popper 1979: 352) in formulating predictions and technical applications for actions which bear on the physical world. Science should thus formulate appropriate directions for actions which intervene in the physical world. This should be done on the basis of the best tested scientific theories. The procedure of prognosis – the prediction of events and states which may/will come to pass at a future date in the physical world if certain conditions are changed through actions – is an inversion of the basic schema of scientific explanation. While an explanation tries to find an appropriate *explanans* for a given *explanandum*, in the case of a prediction the *explanans* (theory from textbooks, primary conditions from observations) is given. However, 'what remain to be found are the logical consequences: certain logical conclusions which are not yet known to us from observation. These are the *predictions*' (Popper 1979: 352). The technical application of the *explanans* (technology), i.e. the use of the *explanans* as the appropriate means for achieving a given goal or aim of action, has the same basic structure as a prediction. The only difference is that, in place of the *explanandum* it is not the prediction but the initial conditions which are given: these are the means which must be available to achieve the goal of the action. 'What are to be found are certain initial conditions which may be realized technically and which are of such a nature that specifications may be deduced from them, together with the theory' (Popper 1979: 353). Thus, from the theories, measures (or means–end relations) are derived which make it possible to achieve the desired goal.

Empirical research into the social world

Social engineering is the primary aim of empirical sociological research (Popper 1979: 94, 349ff.). Social engineering is the process whereby 'social scientists' should propose means to achieve goals which are likely to remove 'problem situations'. Since social problems usually arise out of the unintended consequences of actions, social researchers also have the task of 'analysing the unintended social repercussions of intentional human actions [. . .] and foresee them as far as possible' (1969: 95).

Methodological individualism

This brief outline of empirical social research presupposes an acceptance of the postulates of methodological individualism, and requires the application of the procedure of situational logic, or situational analysis.[16]

Methodological individualism is, according to Agassi (1960: 245ff.), a socio-philosophical approach in opposition to methodological holism (of which Marx was the most important proponent). Other relevant traditions which should be borne in mind are ontological holism (represented by collectivists such as Durkheim and Malinowski) and ontological individualism (with its psychologistic arguments in the tradition of classical and cognitive behaviourism).

The basic postulate of methodological individualism demands

> that all social phenomena, and especially the functioning of all social institutions, should always be understood as resulting from the decisions, actions, attitudes, etc., of human individuals, and that we should never be satisfied by an explanation in terms of so-called 'collectives' (states, nations, race, etc.).
>
> (Popper 1969: 98)

'Methodological individualism' has received a bad press because it is usually confused with ontological individualism. Ontological individualism is summed up perfectly in Margaret Thatcher's statement: 'Society does not exist. There are only individuals.' Because of this confusion, I think a better term for some of what Popper is referring to as 'methodological individualism' is subjective agency. However, in this discussion of Popper, I shall continue to refer to methodological individualism, but stress that this term should not be misunderstood: acceptance of methodological individualism does not imply the rejection of the existence of social collectives and institutions. Neither is it necessary to agree with the statement that a society is no more than the sum of the individuals belonging to it, or that society can be reduced to, or explained in terms of, individual psychology. For 'methodological individualism (implies) not a reduction at all, but a denial of the possibility of attributing aims and hence actions to non-individuals [. . .] [such] as the "economy", "proletariat", "church", "Foreign Office", "industry", etc.' (Jarvie 1972: xiii). From the point of view of methodological individualism, the 'actions *of* groups' are only accessible to social research 'with the aid of the actions of persons *in* groups' (Brodbeck 1968: 283). Like Weber before him (1968: 13ff.), Popper understands collectives as totalities consisting of agents (or subjects), their intentions, decisions, actions and the consequences arising from actions. The consequences should furthermore be regarded as 'more' than the sum of the intentions of the agents' actions. On the one hand, actions relate to each other; and, on the other, the consequences of actions independently affect further actions. These (and like-minded) proponents of 'methodological individualism' are not thereby denying the existence of collectives. But they do maintain that the only meaningful *methodology* of social research lies in the investigation of society through actions.

Methodological individualism, by this definition, means that to a certain extent, social institutions are the product of the intentions of agents and of the consequences of their actions. The adoption of methodological individualism as a *principle* of social research does not mean that agents and their actions are manoeuvred into a social vacuum. An agent's aims can never be free of social influences. But neither can they be determined solely by society or by a 'collective'. People who, by their actions participate in a collective, a social movement, e.g. a union, a regional action group, may certainly all be pursuing the same goal. The communal goal may even in certain circumstances be the sole reason for their association in a group, since every member may have realized that a personal goal can be better and more easily achieved with others. But we should not conclude from this that the group, the movement 'as such' determines the goal of its members. It is at best other subjects, other members of the group or movement, who influence members' goals. In a similar manner, society consists of the actions of the agents who participate in it, and only subjects, not groups or social classes, can have goals. These are the central postulates of 'methodological individualism'.

'Methodological individualism' also holds that all empirically valid macro-analytical statements made by the social sciences must be reducible to true statements about the actions of agents and their consequences. More formally: hypotheses about regular social patterns can only be correct if they correspond to regular actions of agents, with the related (regular) consequences they entail. If this is not the case, macro-analytical statements such as those concerning political economy, structuralist sociology, etc., are at best crude approximations to social reality. But as long as research follows the basic principles of 'methodological individualism', regarding its units as statistical aggregates and not as hypostasized wholes with their own goals, the proponents of methodological individualism have no objections. Indeed, in many instances the aggregation procedure is justified for the simple reason that it is better to achieve a crude approximation than to make no meaningful statement at all. Thus it is appropriate to talk of 'the state' or 'the oil industry', etc., if these terms are meant as a kind of 'abbreviation' for a typical set of goals, actions and consequences of actions of agents in certain social positions.

What is, however, unacceptable from the standpoints of 'methodological individualism' is the assertion that social processes depend solely on 'macroscopic aspects' (Brodbeck 1968: 299), which reductionist Marxism has argued but has been unable to prove empirically. Similarly, the assertion is rejected that social collectives totally determine the goals of agents. In sum, for methodological individualism it is impossible to make prognoses about the development of society from the standpoints of collectives or structures alone. Or to say the same thing in my terms, collectives and structures do not and cannot act without the fuel of subjective agency.

If we accept the postulates of 'methodological individualism' as general research principles, the facts to be explained by the social sciences consist of the actions of agents, the (intended and, more importantly, the unintended) consequences of these, and the problem situations thereby produced. The influence of institutions on the actions of individuals should be included in the explanation, an influence which Popper sees as a possible objective constraint on agents, as are all other macro-aspects of society.

Situational analysis

Critical to Popper's approach is this: explanations should be achieved through the procedure of situational analysis and not by causal explanation. Causal explanation as a concept is sufficiently well known as to mean that no detailed account of it is necessary here. Simply put, it means that A will lead to B. Situational analysis is one of Popper's most important philosophical ideas. It is, however, 'unfortunately nowhere fully explained outside of lectures' (Jarvie 1972: 5), and certain aspects of it remain ambiguous. Not surprisingly, it has given rise to controversial discussion.[17] In my view, the main deficiencies of these debates are, first, that they do not systematically and consistently take account of Popper's epistemology. Second, they are not conducted with reference to the empirical research practice of the social sciences. Third, they leave out entirely the main aim of social research as formulated by Popper: namely, 'social engineering'. In order to help overcome these deficiencies, I have based the following reconstruction and interpretation of Popper's proposals for social scientific methodology on two fundamental assumptions:

1 Situational analysis is a research method which is compatible both with my reconstruction of human actions according to Popper's three-world theory, and with the basic postulates of critical rationalism.
2 Situational analysis is a specific application of Popper's 'biological approach to the third world' (1979: 112). I shall explain this.

The reconstruction of human actions according to the theory of knowledge has already been examined in detail. In the following discussion of situational analysis I shall only refer to it where necessary for clarification. I turn now to an explanation of the idea of a 'biological approach to the third world'.

Popper illustrates the 'biological approach' with the example of an analysis of the behaviour of spiders in relation to their webs mentioned above.[18] In his view this biological approach can also be applied when analysing the way human beings act and the products which result from their actions. In the light of his arachnid example, Popper (1979: 112) requires of social scientific methodology that it should distinguish sharply between two main kinds of problem: 'problems concerned with the acts of production', with the relationship between the producer and the product, and 'problems concerned with the objective structures of these products themselves and their feedback relation' on human actions, with the relationship between the objective structures and the actions.

On this basis Popper (1979: 114) formulates the thesis

> that we should realize that the second category of problems, those concerned with the products in themselves, is in almost every respect more important than the first category, the problems of production

and

> that the problems of the second category are basic for understanding the production problems: contrary to first impressions, we can learn more about production behaviour by studying the products themselves than we can learn about the products by studying production behaviour.

In the light of the idea that the objective structure is the product of actions, we might assume that a research approach along these lines (from the conduct of production to the product) is more scientific because it operates in a more logical way: from an analysis of the reasons for an action to an analysis of its consequences. Popper, however (1979: 115), believes 'that this argument is mistaken', for the following reasons:

> In all sciences, the ordinary approach is from the effects [consequences] to the causes [reasons]. The effect [consequence] raises the problem – the problem to be explained, the *explicandum* – and the scientist tries to solve it by constructing an explanatory hypothesis.
>
> (Popper 1979: 115)

And the hypothesis, we could continue, involves stating a conjectural cause, a conjectural reason for the occurrence of the problematical effect/consequence. This procedure thus corresponds to deductive research logic and should be preferred because of its rational and logical validity.

The method of situational analysis has the same basic structure. Popper (1979: 179) describes it as 'a certain kind of tentative or conjectural explanation of some human action which appeals to the situation in which the agent finds himself'. Also, 'we can try, conjecturally, to give an idealized reconstruction of the *problem situation* in which the agent found himself, and to that extent make the action "understandable" (or "rationally understandable"), that is to say, *adequate to his situation as he saw it*. Situational analysis should thus make possible the explanation of human actions, which in turn should be seen as an attempt to solve problems. The problems to be solved consist in the achievement of goals with the aid of limited means.

The explanation of actions should be concerned with two aspects: first, the aspect of 'logic', i.e. the way in which the agent tries to find out means available within the situation, for the realization of his or her goals; second, the aspect of the 'situation', i.e. the limited means and the constraints which inhibit the realization of the goals. We have now touched on the most important elements of situational analysis: 'goals', 'situation', 'logic' and 'explanation'. I shall now examine more closely their meanings in respect to Popper's line of argument.

Goals

The decisive point in Popper's argument is that the social scientist should not primarily be concerned with goals in explaining actions, as methodological individualism seems to prescribe. He gives two reasons for this:

First, Popper sees 'goals' as mental states, and if we were to consider only the latter we should have to fall back on unacceptable psychologistic arguments. Since he holds that all mental states are shaped by world 3 and therefore accessible to all (i.e. they have objective content), there is no point in taking a detour through the agent's mind. In Popper's view actions are distinguished not so much by their goals but by the situations in which agents attempt to realize their goals. Popper (1969: 96) does admit that some actions can be explained in terms of 'goals' alone. But the explanation can only be successful in those cases where the situation does not thwart attempts to achieve a goal. That is to say, attempts at explanation which relate to goals can only succeed in relatively straightforward cases. And if the job of social scientists is to suggest ways of solving social problems, and as a problem by definition is not straightforward, this methodology is useless: the tasks and the problems of the social sciences begin where established reasons end.

However, Popper (1968: 95) sees the attempt to explain general social problems, such as social injustice, through reference to the goals and intentions of actions, as untenable. This approach leads in his view to the idea of a 'conspiracy theory of society', an idea which characterizes crude Marxist attempts at social explanation. If such explanations were valid, it would mean that social problems were consciously and intentionally caused by a conspiratorial group.

The second argument, as I have already indicated, is that the main aim of social scientists should be to explain social problems resulting from the unintended consequences of actions, and to propose improvements within the field of social engineering. Explanations which relate solely to goals will necessarily fail in this task, for such problems arise from unintended consequences, even though they are the result of intentional actions. If only the causes of actions are cited in explanation, the unintended consequences will be ignored. Therefore, according to Popper, the goals of human actions play only a subordinate role in explanation.

The main points of an explanation should be concerned with the situation of the agent, and the goals should be subsumed under the 'problem situation' of the agent.

Situation[19]

In Popper's view (1979: 179), the social scientist should be concerned primarily with the 'situation' as it is defined by the agent, and with how the agent goes about 'approaching reality'. For 'the action of a human being depends not so much on the situation as it is, but rather on the agent's conception of the situation' (Donagan 1975: 94). The agent's conception of the situation is also part of the situation, a component of the problem situation. It is the task for 'situational analysis to distinguish between the situation as the agent saw it, and the situation as it was (both, of course, conjectured)' (Popper 1979: 179).

As I noted earlier, from the methodological point of view there should be a clear distinction between the physical–material and the social components of a situation. This requirement is particularly important for explanations of the unintended consequences of actions, and for their social engineering. Researchers – especially in the field of ecology – must be familiar with the specific aspects of the 'logic' of connections between the physical and social worlds if they wish to claim empirical validity for their conclusions. In the reconstruction and understanding of the objective situation, and the formulation of problem-solving technologies in the physical–material sphere, researchers should also take account of scientific theories which are at present held to be valid. In the case of the social component, they should take account of sociological theories which are at present held to be valid. Both, the physical and social aspects should be considered in an integrated way, however, if this is necessary for the explanation of a problem situation.

Logic

As we have seen, Popper starts from the idea that when faced with action situations human beings act in a logical way, i.e. rationally. He suggests that the methodology of social research (1967: 144) should also take account of the hypothesis 'that every agent acts adequately or appropriately, that is to say, towards the situation he wishes to achieve' (my translation).[20] Popper terms his thesis of 'action appropriate to the situation' the rationality principle. None the less, Popper (1967: 144) describes the rationality principle as 'of course a more or less empty principle'[21] or, in other words, as a kind of 'nil hypothesis' or 'zero principle', 'which is practically [empirically] invalid and can therefore hardly exclude alternative actions' (Schmid 1979: 498). At the same time, Popper claims that almost every action can be explained by the 'rationality principle'. How are we to interpret these conflicting statements?

The *first* assumption, the formal rationality of humans, 'makes it possible to construct comparatively simple models of their actions and interactions and use these models as approximations' (Popper 1960: 140–1). This idea of 'approximations' leads to the *second*, much stronger assertion, that human beings act in ways appropriate to the situation. If this is a correct interpretation of Popper's ideas, it seems obvious that he is not claiming that human beings always act in a way that is appropriate to the situation. For the second assertion is purely methodological, and should not therefore be interpreted as a statement claiming empirical validity.

In adopting the rationality principle we are in a position to establish whether an empirically observable action is compatible with it or not, and if so to what extent. 'This principle does not have the role of an empirical-explanatory theory, of testable hypothesis' (Popper 1967: 144),[22] as is the case with an empirical causal law. As Weber has pointed out, it should be understood rather as a standard 'of what is to be considered rational in the situation in question' (Popper 1969: 97). The rationality principle should therefore be seen as a kind of 'ideal type'.[23] For the rationality principle, like the ideal type, abstracts 'from certain characteristics and features of a given state of affairs, and tries to establish mentally a smooth process of rational action' (Schmid 1979: 499). The rationality principle should therefore be understood as a casuistic principle, which at best leads to an approximate understanding of a given empirical action.

In the 'method of the logical and rational construction of such models' and their possible application to action situations, Popper (1960: 110) locates the central difference between the natural and the social sciences.

Explanation

The explanation procedure of situational analysis is in Popper's view a special case within the hypothetical-deductive method of explanation in the field of social science. More precisely, it is a 'schema of problem solving by conjecture and refutation, (which) may be used as an explanatory theory of human actions, since we can interpret an action as an attempt to solve a problem' (Popper 1979: 179). This procedure corresponds consistently to Popper's description of human actions in a given situation: sociological explanation is deductive because the logic of the agent's situation is deductive. However, and this brings us to the main point, with this suggestion Popper is neither following or endorsing those action-science methodologists who argue along causal lines, despite the fact that they claim Popper's authority.[24] He expressly describes the 'sociological explanation as a rational explanation' (Schmid 1979: 492) and not as a causal explanation: sociological explanation cannot be causal because the logic of the agent's situation is not deterministic in the causal sense. And the logic of the situation cannot be causal because all the (natural and social) factors of the situation can only be necessary, but never sufficient conditions of action. This means that an agent needs certain situational conditions to complete the intended action, but the fact that these conditions are given is never a sufficient guarantee that the action will be carried out.

The result of a model construction has the same function in sociological explanations as causal laws have for explanations in the natural sciences. Popper (1967: 143) sees model construction as 'the only means to reach an adequate explanation and understanding of social processes'.[25] Admittedly, these models enable us to find fewer 'detailed explanation' ['explications en détail'] than the natural sciences, but on the other hand they enable us to find 'explications in principle' ['explications en principe']. In effect, Popper is proposing that in our explanations of social actions we should replace causal laws by model constructions. The proposition is inherent in his basic conviction that human social activity is not determined in the causal sense, but is directed by deliberation, goals and social rules, all of which can be influenced by rational argument.

As far as the practical application of this procedure in research is concerned, it must be remembered that for Popper the main aim of social research is critical 'social engineering'. The first condition for the achievement of this objective is that we be in a position to understand (approximately) the agent's definition of the situation. Applying the rationality principle we can then produce an initial hypothetical explanation, which can subsequently be tested with reference to the actual course of an action.

The *explanandum* of an explanation is thus an action which has a particular problematical, usually unintended, consequence. The *explanans* consists, with the inclusion of the rationality principle, of statements about the physical–material and social factors of the situation, and also the agent's knowledge about the situation. The *conclusion* that can be reached is this: as a result of the (objective) situation and its assessment by the agent, 'these' consequences have occurred/will occur. The following example (which is not my own) serves to illustrate the procedure:

> *Explanandum.* How can it happen that Mr X, an otherwise careful driver who has never had an accident, suddenly causes a pile-up?
>
> *Analysis of the situation.* The accident occurred on a motorway. Mr X did not realize that the traffic *situation* was quite different from that on a typical urban main road. He acted on the motorway according to the logic of urban driving.
>
> *Conclusion.* Although it was Mr X's aim to drive just as carefully on the motorway as he would in town (assessment of the situation), in the 'motorway' situation this led to a pile-up (unintended consequence of an action).[26]

Most people on the motorway are pursuing the same goal (to get from A to B quickly and safely), and therefore their situation is, objectively speaking, identical. Their actions, however, may be confronted with specific problems. Following Popper's line of thought, we could say that where agents do not adapt their logic appropriately to the 'motorway' situation, special problems arise as the unintended consequences of the building of motorways. Such problems result from the facts of the situation, and their assessment by agents within the framework of a certain action.

It should be clear from the above that it is only to a very limited degree that Popper's social explanation implies any 'law', i.e. a statement about a regular action pattern. His form of social explanation is applicable to both regularly recurring and unique modes of action. Where actions

> are regular and repeatable, we search for laws [such as]: 'All social changes create vested interests which resist further social changes [. . .]'. Where an action is of a rare or unique kind, we seek no sociological laws of it; we simply see how close an approximation to it we can deduce from the assumption of rationality.
>
> (Jarvie 1972: 18)

In such cases the application of ideal types should be interpreted in a deductive sense. It is a heuristic principle taking the form of bold conjecture and empirical refutation. The closeness of the approximation can be measured by a comparison of the hypothetical deduction with the actual facts of the case. If they coincide, a provisional explanation has been reached which must be subjected to further tests. If they diverge, the procedure of situational analysis must be continued until a satisfactory approximation has been reached.

How should we proceed in our empirical research if we wish to use situational analysis to explain actions or eliminate social problems arising from unintended consequences? I propose that the following steps be considered a provisional outline for empirical research, which of course need further discussion and practical testing.[27] The first step is the *explanation*, which in turn has four sub-stages.

Stage 1.   Empirically valid formulation of the *explanandum*.
Stage 2.   All mental aspects must be replaced by the basic thesis of the theory of knowledge: formal rationality, and by the rationality principle. The two together make possible

Stage 3. 'Idealized reconstruction of the problem situation in which the agent found himself' (Popper 1979: 179), i.e. construction of a model of the problem situation.

Stage 4. Conclusion from the situation about the mode of action and/or the unintended consequence of the action.

The second step bears on *social engineering*. The four sub-steps outlined above should also be used in the formulation of a social engineering statement, but such a statement requires three additional sub-steps:

Stage 5. Empirical inquiry into the objective structure of the physical–material and social aspects of the problem situation.

Stage 6. Reference to the scientific knowledge relevant to the problem, from which an empirically and rationally valid procedure for the situation can be derived, meaning deduction of the most effective means for achieving the goal.

Stage 7. Formulation of directions for actions intended to eliminate or avoid the social or ecological problem.

SUMMARY

It is important to emphasize that, in Popper's explanation of human social actions, 'goals' are not necessary in his argument. 'Goals' should be subordinated to the situation overall. In attempts to explain human actions, causal laws are replaced by the rationality principle. Popper is prepared to sacrifice causal explanation in order to avoid the 'nightmare' of determinism.[28]

In social explanations, Popper's emphasis is precisely on the situation and conceptual approximations to it, rather than on causal predictions. This is contrary to the received view of Popper's 'social' epistemology. Popper does not argue that the methodology of the social sciences should follow the 'if/then' causal model he outlines for the natural sciences. To make the point more strongly: the structural-functionalists, positivist sociologists, and neo-determinist geographers, who claim that social science research should be based on 'Popper's epistemology', are invoking a Popper who does not exist (a striking, unintended consequence of Popper's own intentions).

Chapter 3

The subjective standpoint

We have seen that K.R. Popper's philosophy of science has as its starting point an 'objective theory of knowledge without a knowing subject'. Edmund Husserl, the founder of phenomenology, and Alfred Schutz, phenomenology's most important if under-appreciated representative in the context of the social sciences, start from a diametrically opposed thesis.[1] Their philosophy proposes that the conditions of objective knowledge reside in the knowing subject. There is no knowledge independent of the subject, for everything knowable must first be analysed in terms of what constitutes its meanings.

In the following I shall be examining this 'subjective perspective' thesis. I shall also try to give a systematic account of its significance for the social sciences and for social geography. My analysis focuses mainly on the work of Schutz, and I shall only refer to Husserl for the sake of clarity. My prime concern is with general principles of theories of knowledge and science, and the norms derived from them for social research. I shall concentrate especially on the phenomenological theory of the constitution of the meaning-content of social action. But I shall first outline the historical context of the central theses of the phenomenological theory of knowledge. This will facilitate an understanding of the subjective perspective in cultural and social research.

THE SUBJECTIVE PERSPECTIVE'S HISTORICAL CONTEXT

Galileo's laying of the foundations of scientific thought led, in epistemology, to a dualism which found its most consistent expression in the work of Descartes. This propounded a duality of the physical–material (external) world and human consciousness (internal world). Knowledge became the problem of mediation between the external world (body) and the internal world (soul). Hume and the British school of empiricists proceeded to analyse the part of the external world that exists in human consciousness, how it gets there and how it is 'processed' into knowledge. They concluded that no knowledge about the external world can be anything other than a belief. As I discussed in the last chapter, Hume also concluded that this belief becomes all the firmer, the more often our consciousness receives certain sense impressions from outside. Thus any experience of reality can only lead to a degree of certainty about that reality's probability. This idea led at the end of the nineteenth century to the speculative approach of Hegel's idealism and Dilthey's and others' neo-idealism (cf. Küng 1982: 2).

Hegel may have been happy that this synthetic philosophy of pure speculation in turn produced its own antithesis: the analytical school (Frege, Russell, Carnap, Popper and others). Speculative philosophy also led to another reaction. This is the pheno-menological school (Brentano, Husserl, Scheler, Merleau-Ponty, Heidegger and others). These two schools are in agreement on the need for an exact analysis of reality, rather than (speculative) global syntheses. Both want a 'philosophy of rigorous science' (Husserl 1965). The two poles differ, however, in their responses to Cartesian dualism. The analytical school is so steeped in the idea of an external world independent of human beings, that it believes it can use that external world as a criterion of judgement to test the truth of linguistic conventions (by which it means defined concepts) and scientifically formulated laws. The meaning struc-ture of the 'real world' is – in this sense – taken (uncritically) as given. From the standpoint of the phenomenological theory of knowledge, this is tantamount to an objective resolution of Cartesian dualism. The phenomenological school tries to render this dualism invalid by taking the consciousness of the agent as its starting point. It thus reverts to the position of Descartes, though with a different emphasis.

The main concern of Husserlian phenomenology is to develop the foundations of a 'rigorous science', although this is not the same as the demand for mathematical formalization, or the con-viction that the only true knowledge is measurable knowledge.[2] As a trained mathematician who wrote several major works on mathe-matical logic,[3] Husserl was convinced

that none of the so-called rigorous sciences, which use mathematical language with such efficiency, can lead toward an
understanding of our experiences of the world – a world the
existence of which they uncritically presuppose, and which they
pretend to measure by yardsticks and pointers on the scale of
their instruments. All empirical sciences refer to the world as
pre-given; but they and their instruments are themselves
elements of this world.

(Schutz 1962: 100)

With his demand for a 'rigorous science' Husserl had a more
comprehensive programme in mind, one that went beyond the
formalization of scientific statements. He wanted to lay bare all the
preconditions of knowledge on which both the natural and the
social sciences are based. To this end he attempted to formulate
a theory of knowledge which would uncover the hidden
preconditions of all habitual thought, so that objectively true
statements could be made.

The 'merger of rationalism and empiricism, from which was
fashioned the hard core of modern science' (Luckmann 1983: 22),
led, for Husserl (1970a), to the 'crisis' of European science. It
developed, in his view, out of the alienation of the idealized and
formalized products of theoretical activity in the fields of mathematics and logic. These were not only alienated from their roots in
the life-world. They were also reified as structural principles of
nature. Husserl consequently reproaches them with 'blind
naivety', in the sense that they were blind to the roots, the bases of
their 'theoretical activities of idealization and mathematization,
and the foundation of these activities in the *praxis* of everyday life'
(Luckmann 1983: 18). Luckmann remarks further that the social
sciences have also been prone to this blindness, but that their
scientific ideal is characterized by a further naivety.

Social science not only very properly took over the logical form
of reasoning (the historic achievement of the combination of
empiricism with rationalism) from the physical science; it also
very improperly pretends to the same (illegitimate) *epistemological* autonomy of scientific knowledge.

(Luckmann 1983: 18)

Thus in traditional empirical social research the 'blindness concerning the nature of the subject matter of social science [. . .]

consists in the metaphysical elimination of reflexivity' (Luckmann 1983: 18) and of intentionality.

With his studies of intentionality, and his constitutional analysis of consciousness and meaning-contents and of the inter-subjectivity of the knowing subject's cognition processes, Husserl pointed to the central aspects necessary to counteract this aliena-tion. Every scientific analysis, in the phenomenologists' view, should be anchored in the life-world, so that the alienation which has developed in the field of science can be eliminated. This blind naivety can only be eliminated if it is borne in mind, even in scientific actions, that every act of acquired knowledge is founded on the stream of consciousness of the knowing subject, and that the consciousness is based in the life-world.

Husserl is especially motivated by the fundamental tension that exists between the various demands made on 'truth'.

> A true discovery must fulfil two requirements, which at first appear contradictory: on the one hand it has to be *'objective'*: the content of the knowledge must be valid independently of the subjective circumstances in which the discovery was actually made. On the other hand, however, we expect the knowing subject to have convinced himself in some way of the validity of his discovery, and this after all he could only have done by making the discovery *subjectively* in a concrete situation.
>
> (Held 1981: 275)

In other words, the 'objective' can only ever be the product of a subjective act, but can nevertheless exist independently of it. If we try to resolve the dilemma from a one-sided objective viewpoint, we lapse into a naive-positivism which uncritically takes reality as the criterion of judgement for objectivity. If we resolve it subjectively, we lapse into an (idealistic) psychologism which assumes that logic-ally valid propositions are psychological, and not objective facts.

Despite the refutation of psychologism in his *Logical Investi-gations*, Husserl's resolution of this dilemma turns to the subjective side of understanding. He starts out with the idea that what has been recognized as true is 'objective' or valid 'in itself'. That is to say, it does have an existence independent of a change in the subjective situation of actual discovery. But the content of truth which resides precisely *in* that independence is only achieved through its unique relation to subjective knowledge. Husserl thus declines to accept the realism postulate uncritically, at the same

time resolving not to become trapped in metaphysics. He therefore no longer seeks the causes of what exists in the world and what
takes place in it, but the constitution of the *meaning* of the facts to
be discovered.

This notion involves a radical attempt to question critically all
experience data which seem to be simply 'given'. To this end the
experiencing subject must be freed from the 'naive' approach, an
idea that Descartes also took as his starting point. Descartes no
longer regarded sense data as the origin of experience, but proposed the *ego cogito*, the thinking subject, as the origin of our
knowledge. This is also Husserl's point of departure, and to that
extent his theory of knowledge can be described as a subjective
theory of knowledge. Husserl's criticism is, however, directed
precisely at these two components of the Cartesian theory of
knowledge: too little importance was given by Descartes to both
the concept of *ego* and the concept of 'cogitation'. Descartes
regarded acts of thinking ('cogitations') which occur in the ego's
consciousness as isolated wholes. He did not see single acts of
thinking in relation to prior acts of thought. Husserl also accuses
Descartes of not having made 'a sufficiently radical distinction
between the act of thinking and the object of thought' (Schutz
1962: 102). Hence Husserl's own thought, and radicalization of
Descartes, is characterized by his use of the term 'intentionality of
thinking'.

In contrast to Descartes, phenomenologists emphasize the
intentional character of all our thinking. For them there are – also in
contrast to Frege and Popper – no thoughts, or ideas as such. For
thoughts always refer to things that are experienced, so that every
act of thinking, remembering, imagining, etc., must be a thinking
of, remembering *of* or imagining *of* the object that is thought,
remembered or imagined. Every thought act thus has an intentional structure. With the discovery of the intentional structure of
consciousness it is possible to resolve the dilemma of the seemingly
contradictory demands on 'truth', in a way that is neither one-
sidedly 'objective' nor one-sidedly 'subjective'. 'Truth' in this context means a set of facts which is restricted neither to the external
world nor to psychological factors, but is the product of an act of
consciousness which combines the two components and thus by-
passes Cartesian dualism. With intentionality we can concentrate
on the 'object' without neglecting the subjective act of knowledge.
The phenomenologist maxim *'towards the things themselves'* should

also be understood in this light. When the theme, 'consciousness-of-things', is made central, and 'consciousness' is thereby indissolubly linked to 'things', the subjective is similarly inextricably tied to the external object.[4] With this approach the postulation of an idealistic or solipsistic view of the world is just as impossible as the 'objectivist' elimination of the subject.

Husserl's second radicalization of the Cartesian theory of knowledge in relation to acts of thinking concerns the distinction between that which is perceived and the intentional act of perceiving. Although the perceived object exists in the external world independently of the perceiving subject, the latter can only perceive it as it appears to him or her and not as it actually exists 'as such' (whatever that may mean). The perceived phenomenon (as the product of an intentional act of consciousness) is thereby also independent of that which is 'actually given'. Whatever may happen to it in the external world later on, it can continue to exist in the subject's consciousness just as the subject perceived it at the time, independently of its actual external fate. If the subject becomes concerned with the same object at a later date, he or she may remain true to the original perception (because he or she has not changed, and because the intentional act of consciousness has remained consistent). If this is the case, the consciousness may achieve a synthesis, a recognition of the two phenomena as identical. The second perception may, however, prove to be inconsistent with the first. Then '(ego) may doubt either of them, or search for an explanation of their apparent inconsistency' (Schutz 1962: 107). If ego decides on a clarification of this doubt, there must be an initial clear distinction between the act of perceiving and that which is perceived.

Husserl gives the name '*noesis*' to the act of perceiving, and calls the perceived '*noema*'.

> There are (in this context) modifications of the intentional object, which are due to activities of the mind and are therefore noetical, and others which originate within the intentional object itself and are therefore noematical.
>
> (Schutz 1962: 107)

Every perception (*noesis*) is thus the product of a complicated process of interpretation 'in which the present perception was connected with previously experienced perceptions (*cogitationes*) of the different aspects' (Schutz 1962: 108). All earlier experiences have created a certain universal style, a certain construct of type-structure. All later perceptions are interpreted in relation to this type-structure. In other words, the noematic correlate of the *noesis* is given its particular meaning by the intentional act of consciousness, which in turn refers back to a type-structure already present in the ego's consciousness. In order to find out whether the inconsistency of a present experience, compared to an earlier one, is conditional upon the noetic (act) or the noematic (perceived) aspect of knowledge, we need to turn to Husserl's discussion of the role of the ego in the process of understanding.

In order to assess the important role of the subject, i.e. the ego, in the process of knowledge acquisition, an uncritical, naive attitude to the world has to be given up. For in the (naive) approach to knowledge we are not interested in whether the world is just as we perceive it, or whether its mode of existence depends solely on our perceptions. In order to be sure of the significance and nature of the contribution of the subjective component of knowledge, we need, to a certain extent, to have an 'artificial, a theoretical attitude'.[5] Phenomenological philosophy and phenomenologically oriented social science have set themselves the task of analysing and classifying such attitudes.

CONSTITUTING PROCESSES AND INTERSUBJECTIVITY

For the proponents of the phenomenological perspective, the objective meaning of social and physical reality is not pre-given as such. It should rather be seen as the product of a meaningful construction by interacting subjects. As one phenomenologist has it, 'the objective world is that constituted by everyone, and everyone is (in order to have the same experiences and acquire the same knowledge) interchangeable with every other similarly constituted alter-ego' (Szilasi 1959: 111). 'Objective meaning' is thus the communal and associative relationship between the subjects' streams of consciousness, i.e. their 'stock of knowledge at hand'. By intentional acts, this communal and associative relationship constitutes an intersubjective reality.

I shall now examine the main concepts of this theory of knowledge in more detail, in order to clarify the basic thesis just outlined.

The term '*constitution*' is used in phenomenological literature in two different contexts: first, in relation to the constitution of the knowing subject's consciousness, and second, in relation to the constitution of the meaning-content of the phenomena perceived. On the first aspect of 'constitution': we must start out from the fact that any understanding of the meaning of a set of circumstances points back to other, earlier intentional experiences. These may be the agent's own experiences or those of others which have been passed on to him. Whatever the case: every act of understanding, phenomenologists argue, presupposes a certain prior knowledge about the set of circumstances in question. According to this theory, consciousness, Schutz's 'stock of knowledge at hand', is the sediment of all previous experiences which at the time of the particular experience or action lie in the past. The knowing subject can always bring them back to mind, however, as they belong to the 'stream of consciousness'. Thus, during the course of the knowing subject's life, a certain structure of consciousness structure is built up ('constituted'), which Schutz analyses in detail in his work on relevance structure and ideal types.[6] If subjects make their different experiences in the same culture and society, the consciousness of each individual in that community will (more easily) attain an intersubjectively consistent structure. Thus in the stream of lived experience, in the subject's stock of knowledge, the preconditions are created for the constitution of 'objectivity': one given event or object will have the same meaning for everyone, thereby becoming 'objective' fact. In this way an intersubjective consciousness *of* an object is constituted. In order to describe a truth as 'objective truth', it must first be shown that a given fact [*Sachverhalt*] 'appears the same to myself and to all those with whom I communicate' (Brauner 1978: 42). Objective truth is only possible if a given fact, to which a statement refers, is acknowledged to be identical by all subjects who are in communication with each other.

An intersubjective consensus on the world is therefore possible, despite the decisive role played by the subject in the process of cognition. This is explained by the idea that the subjectivity of the ego is merely one pole in the constitution of consciousness. The second pole is precisely the intersubjectivity of the world of contemporaries. Through socialization processes the knowing subject learns to interpret the world in an intersubjective way; the central element here is learning a common language. In these socialization processes, 'objective' perspectives of cognition are conveyed, so that given facts with the same meaning can be shared by all subjects with the same socio-cultural world. If the perspectives are mutual, the facts become objective, i.e. intersubjectively identical, object of knowledge. This means, as Brauner puts it, 'that although the world with its "things" enters the consciousness (of every knowing subject) in various ways, it can nevertheless be experienced intersubjectively as a world common to all, by means of acts of cognition and appearance' (Brauner 1978: 42). Because of the common perspective, individual experience of the world fuses into one single intersubjective world, culminating in 'common objects of consciousness' (Husserl 1982: 56): a world that appears objectively given.

The second aspect of 'constitution' is the subjective constitution of the meaning of physical–material and mental–immaterial facts. This is an intentional endowment of meaning, based on the intersubjective stock of knowledge. 'Intentionality', in this context, is both a particular direction of consciousness and an endowment of meaning. Husserl is here primarily interested in the creative function of intentionality inherent in the act of perceiving. During this act the sense data (light impulses, sound waves, etc.), Husserl's 'substance' (*Hyle*), become a *noema*, a phenomenon with a specific meaning-content. From the phenomenologist point of view, pure sense data, on which the classical positivists base their objective science, only become meaningful through the intentional act of consciousness. Thus it is only after the pure sense data have been given 'shape and form' (Peursen 1969: 95) by consciousness, and its meaning constituted as the product of the ego's subjective act of cognition, that they become relevant to the agent.

Although it is the knowing subject who performs the intentional constituting act, this epistemology cannot be described (as Habermas does (1988)) as an egological approach. Certainly Husserl locates the synthesizing process of intentionality in the subject's stream of consciousness. But at the same time it should be understood as related to something other than the subject. Thus the internal and the external, the subject and the world, the agent and the social are brought into coexistence through the act of cognition. The bearing of this approach on social research should be beginning to emerge, which makes this an appropriate point to turn to Schutz.

For Schutz (1962: 116ff.) the reference [*Verweis*] to intentionality, to the intersubjectivity of the cognition process, forms the fundamental basis of the 'social'. When it comes to the distinction between 'things in the "outer world"' (Schutz, 1962: 110) and 'ideal objects', Schutz, like Husserl, starts from the premise that every thing, every given fact in the world can, as an object of knowledge, only be an ideal, intentional object. The real object is conveyed to the consciousness through signs and symbols 'which are in turn perceptible [*wahrnehmbare*] things, such as the sound waves of the spoken word, or printed letters' (Schutz 1962: 110). The conclusion is drawn from this that the real world can indeed exist without a knowing subject, but its meaning-content can only be constituted by the knowing subject. The real world as such is therefore a precondition for the subject's actions, but it only becomes relevant if it takes on a particular meaning, when the subject has constituted its meaning-contents. Thus the construction of social reality takes place through the symbolization of what is knowable in the external world. In this process both forms or aspects of constitution are combined. Any constitution of the meaning-content of a given fact through an intentional act (the second aspect) always involves its own history, 'and this history (the first aspect) can be found by questioning it' (Schutz 1962: 111).

In its own way phenomenological epistemology is not only concerned to show the logical validity of human cognition processes: it also seeks to elucidate the preconditions of formal logic. But to do this, it has to answer the question of which constituting processes earn the judgements 'logically valid' or 'logically invalid'. For Husserl, the 'truth' of a statement or insight can only be assessed when these basic acts of consciousness have been explicated [*geklärt*]. The naive assumptions of objective science about the ontological preconditions of 'reality' are unacceptable. The preconditions themselves must first be analysed.

Phenomenological philosophy aims therefore to investigate what has been taken for granted by scientists: 'The phenomenologist, we may say, does not have to do with the objects themselves; he is interested in their *meaning*, as it is constituted by the activities of our mind' (Schutz 1962: 115). 'Objective reality' is not taken for granted, but classified as the product of the knowing subject's constituting processes. Such constituting processes should be made accessible to empirical research, and not classed as *a priori* categories or as belonging to the realm of metaphysics. I shall attempt to show that this affects the central concerns of all social sciences which have as their objective the formulation of empirically valid statements about social reality.

THE LIFE-WORLD AS AN ESSENTIAL PRECONDITION OF RESEARCH

The basis for the constitution of consciousness is described by Husserl as the 'life-world'. Schutz also calls it the 'everyday' or 'common-sense world of the agent'.[7] The everyday world is experienced by the agent in a 'natural attitude', and according to Schutz it should be investigated by the sociologist in a 'theoretical attitude' appropriate to its meaning. In this section I shall be dealing primarily with this thesis. First the term 'attitude' must be explained before I can examine in more detail the difference between intentional constituting processes conducted in a 'natural attitude', and those conducted in a 'theoretical attitude', and their relationship to each other in Schutz's philosophy.[8]

Different 'attitudes'

The term 'attitude', in Husserl's sense (1970a: 280), denotes 'a habitually fixed style of willing life comprising directions of the will or interests that are prescribed by this style, comprising the ultimate ends, the cultural accomplishments whose total style is thereby determined'. To put it more simply, he understands by this term a 'perspective' which determines the mode of experience. Schutz (1962: 207ff.) uses the term '*attention à la vie*' for 'attitude', an expression taken from Bergson. It means something similar to 'tension of the consciousness', a particular orientation towards and attention to life through which different realms of reality are created.

Following William James, Schutz (1962: 207) sees 'reality' as

> simply [a] relation to our emotional and active life. The origin of all reality is subjective; whatever excites and stimulates our interest is real. To call a thing real means that this thing stands in a certain relation to ourselves.

The term 'real' is thus understood as the 'horizon of meaning'. For agents, 'reality' has objective meaning only as long as there is intersubjective agreement on the existence of a thing, and as long as no contradiction arises.[9]

The various degrees of tension of our consciousness, the styles of experience and cognition, are 'functions of our varying interest in life' (Schutz 1962: 212). According to the level of interest, different tensions of consciousness and correspondingly different realms of reality can be distinguished, 'each with its own special and separate style of existence' (Schutz 1962: 207), thereby forming a finite area of meaning. These classifications of reality are not established through any ontological structure of their objects, but rather constituted through the meaning of the experience of the subject. In other words,

> all experiences that belong to a finite province of meaning point to a particular style of lived experience – viz., a cognitive style. In regard to this style, they are all in mutual harmony and are compatible with one another.
>
> (Schutz and Luckmann 1974: 23).

Schutz classifies the realms of reality in this way: the everyday world, the world of science and research, the world of religious experience, and the world of fantasy (play, art, fairy-tales, jokes, . . .), dreams and madness. For the purposes of social sciences, the everyday world and the world of science and research are of particular relevance for Schutz, together with their respective styles of cognition, i.e. attitudes. Also the problem of (social) science's 'adequate' reference to the everyday world is of central interest. Following Husserl, Schutz calls the style of cognition which constitutes the reality of the everyday world the 'natural attitude', and that which does the same in the world of science the 'theoretical attitude'.

He classifies every 'attitude' more precisely according to its primary relevance, its primary form of spontaneity and its particular form of 'bracketing'. As we shall see below, the primary form of spontaneity in both the natural and the theoretical attitudes is characterized by bodily movements as 'acts of operating' in the outer world, the latter by thought acts.

The natural attitude and the everyday world

Husserl defines the 'natural attitude' as 'immersion in the world' ['*in-die-Welt-hinein-leben*'], where the knowing and acting person is always directed by *pragmatic motives*. It denotes a naive attitude to the world, where the question is not asked whether the world is in fact just as it is experienced or whether, on the other hand, its mode of existence depends solely on the subject's perception. With that the form of 'bracketing' which an agent, according to Schutz, applies in his natural attitude is already delineated in its most general sense: 'What he "puts in brackets" is the doubt that the world and its objects might be otherwise than it appears to him' (Schutz 1962: 229). He takes the world as he constitutes it for granted, and from his point of view as real, for at least as long as it is not called into question and does not become problematical. The natural attitude thus directs my experience of the world into which I was born and which I assume existed before me. 'It is the unexamined ground of everything given in my experience, as it were, the taken-for-granted frame in which all the problems which I must overcome are placed' (Schutz and Luckmann 1974: 4).[10]

The natural attitude's realm of reality comprises both physical and ideal objects. Their meanings are constituted and taken for granted by knowing subjects who are members of a given society and culture. This realm of reality comprises primarily the world of sense-data, the world that is *directly* accessible to perception,[11] where the agent's body represents the zero point of orientation.

The everyday world as an unquestioned basis, constituted within the natural attitude, thus forms that realm or region of reality in which humans participate by means of recurrent regular acts of intervention. In it everything is provisionally accepted without question, being from the outset intersubjective in character. This feature is, however, always experienced by agents through subjective interpretations of meaning, owing to the stock of knowledge which they have acquired earlier. This realm of reality forms the scheme of reference for the acts with which the knowing ego's intentional consciousness elucidates the everyday world. It is made up of ideal-typical knowledge. It contains 'open horizons of anticipated similar experiences' (Schutz 1962: 7) which have already occurred. The world is not experienced as 'an arrangement of individual unique objects, dispersed in space and time' (Schutz 1962: 7), but in its ideal types. 'What is experienced in the actual perception of an object is apperceptively[12] transferred to any other similar object, perceived merely as to its type'[13] (Schutz 1962: 8). In the natural attitude, the agent perceives the life-world in its typical aspects, and within the context of the type-structure, i.e. frame of reference, the agent orients himself or herself accordingly.

In accordance with the pragmatic motives involved, agents in their natural attitude are primarily concerned only with those objects which stand out. They only stand out against a background which is unquestioned. This selective activity determines the particular characteristics of an object which are individual, and those which are typical (cf. Schutz 1962: 9–10). What kind of determination this will be depends on the nature of the stock of knowledge at hand at any one time. Every elucidation of the life-world thus takes place in a biographically determined situation. The biographically determined situation14 is, however, not only determined by the type-structure, but it also makes possible future activity and available goals, possible pragmatic motives. And this 'purpose at hand [. . .] defines those elements among all others contained in such a situation which are relevant for this purpose' (Schutz 1962: 9).

Thus, in the everyday world agents adopt the 'natural attitude' as described above, and their stock of knowledge is formed by means of intersubjective types. For Schutz, however, this is too imprecise a description of the preconditions for the ego's authentic communication with others. According to him, the 'natural attitude' implies a further bracketing: the 'assumptions of the constancy of the world's structure' (Schutz and Luckmann 1974: 7).

Ideal-type knowledge, as constituted in the agent's stock of knowledge and applied in his elucidation of the world, remains unquestioned in the 'bracketing of the natural attitude'. 'Every explication within the life-world (also remains) within the milieu of affairs which have already been explicated, within a reality that is fundamentally and typically familiar' (Schutz and Luckmann 1974: 7). In order to cope with the world, the agent relies on the idea that the world 'will remain further (just as it has been known by him up until now), and that consequently the stock of knowledge obtained from his fellow-men and formed from (his) own experience will continue to preserve its fundamental validity' (Schutz and Luckmann 1974: 7). Following Husserl (1969: 98), Schutz calls this assumption the idealization of the 'And So On'. It comprises the first aspect of the 'assumption of the constancy of the world's structure'.

This leads to another basic assumption. According to Schutz, agents start out with the idea that they can always repeat earlier successful acts. As long as the structure of the life-world remains constant, they can count on being able to intervene in it again in the future, just as they did previously. Schutz and Husserl call this assumption (which is a form of idealization) the second aspect of the constancy of the world's structure, the ideality of the 'I-can-always-do-it-again'.

To summarize: the natural attitude, which constitutes the everyday world of agents, is characterized by its pragmatic 'turning-towards', orientation and attention. In addition, Schutz distinguishes a cultural layer of meaning in the everyday world, 'that first makes physical objects into objects of naive experience' (Schutz and Luckmann 1974: 21). He also singles out agents' relationships and interactions in the natural attitude, the everyday social world. These realms or provinces of meaning are all characterized by a uniform style of experience and cognition, which makes them the finite meaning area of the everyday world. It is the style of the specific tension of the consciousness as a function of pragmatic motives which is the regulatory principle for experience and cognition, namely 'harmony and compatibility restricted to a given province of meaning' (Schutz and Luckmann 1974: 24).

The move from one finite province of meaning into another is described by Schutz as a 'leap' or a 'shock',[15] as the 'exchange of one style of lived experience for another' (Schutz and Luckmann 1974: 24). Every province of meaning is characterized by its own compatibility structure of what is known and experienced, so that one area of meaning cannot be arbitrarily and indiscriminately reduced to another.

Appropriately, for the social sciences, the central methodological problem is the 'leap' from the experience/cognition style of the natural attitude of the everyday world into a theoretical attitude. How can the social sciences adequately analyse the everyday world? Or: what are the conditions for a social science which proposes to analyse society from the subjective perspective of agents, and what procedural methods should/must it apply in order to achieve this goal? Before I can examine these questions I must first analyse the distinctive features of the theoretical attitude. Then I can consider its relations with the everyday world, or the 'leap' from the 'natural attitude' to the 'theoretical attitude', and vice versa.

The theoretical attitude and the everyday world

The 'theoretical attitude', 'the tension of consciousness' in research, differs from the 'natural attitude' primarily in that the scientist who adopts it is *not* motivated by pragmatic considerations. The knowing subject takes up the position of an disinterested observer[16] who 'brackets' any practical personal interests. The objective of the scientist 'is not to master the world but to observe and possibly to understand it' (Schutz 1962: 245), and to formulate a scientific theory about it. Schutz thus accepts Weber's 'value-free postulate' (1951: 475ff.), although he embeds it in his own phenomenological context.

Of course theories can, in their practical application, lead to a better mastery of the world – to the invention of technological innovations, etc. But according to Schutz this application is 'not an element of the process of scientific theorizing itself. Scientific theorizing is one thing, dealing with science in the world of *working* is another' (Schutz 1962: 246). So the issue that concerns us now is not the application of scientific theory, but the formation of scientific theory itself.

As the 'theoretical attitude' implies the 'bracketing' of pragmatic motives, it presupposes that the researcher distances himself or herself from subjective interests. A researcher's natural attitude must be replaced by that of the pertinent discipline. The theoretical attitude should be determined by two things: (i) the immediate problem confronted and a theory appropriated to it; (ii) (related) the background which has already been constituted within the framework of the discipline. This theoretical attitude implies an interest in problems which are valid for everyone: in 'problems in itself'.[17] Researchers also 'bracket' their own physical existence, i.e. they suspend the salience of their own particular (spatial, fixed) circumstances, and live in *ways* of thinking. What this means for the subjectivity of a researcher who truly makes the leap from a natural attitude to a theoretical attitude is another question. For purposes of my subsequent argument, however, it should also be noted that, since in this respect researchers do not participate in the external world, they can continually revise and reject their theories without directly bringing about change in the non-scientific world.[18]

'This theoretical universe of a specialist science is itself a finite province of meaning, having its peculiar cognitive style with peculiar implications of problems and horizons to be explicated' (Schutz 1962: 250). From this premise Schutz derives the rules for scientific action, which can be summarized in the following three postulates:

1 *The postulate of agreement and compatibility of all statements.* This requirement relates not only to statements within one's own discipline, but to all other scientific statements, and to the natural attitude to the everyday world.

2 *The postulate of the empirical foundation for scientific statements.* This means 'that all scientific thought has to be derived, directly or indirectly, from tested observations' (Schutz 1962: 251). That is, these statements must relate to facts within the life-world.

3 *The 'postulate of the highest possible clarity and distinctness of all terms and notions used'* (Schutz 1962: 251). This requirement indicates that, through elucidation within the framework of the 'theoretical attitude', diffuse proto-scientific notions and ways of thinking are moved to a level of higher clarity, in that their (hidden) implications of meaning are made manifest.

From these three general rules, which establish the conditions upon which the meaning area of 'science' can be rendered 'real', Schutz derives fairly precise guidelines for sociological methodology. He first discusses the main objectives of any scientific action, then goes on to compare sociological procedures with the methodology of the natural sciences, and so establishes the particular characteristics of the former.[19]

Following Whitehead, Schutz sees the prime objectives of any science as, first, the development of empirical theories (i.e. theories which tally with experience) and, second, the explication in outline of everyday concepts of the world. This explication 'consists in the preservation of these concepts in a scientific theory of harmonized thought' (Schutz 1962: 4). To this end, science must 'develop devices by which the thought objects of common-sense experience are superseded by the thought objects of science' (Schutz 1962: 4). It is in this process of supersession that we see the meaning of the differences between the natural and the social sciences.

All knowledge of the world, whether everyday, social or scientific in nature, to do with the natural or the social sciences, 'involves constructs, i.e. a set of abstractions, generalizations, formalizations, idealizations specific to the respective level of thought organization' (Schutz 1962: 5). Thus all classifiable facts should be understood as interpreted facts: such interpretations should be seen 'only' as specific aspects of that which we experience. If science replaces constructs of common sense with clearer theoretical constructs, the *natural sciences* then relate to the physical world. This physical world comprises only those facts which have no relevance- and meaning-structure in themselves. Consequently natural scientists are only committed to the relevance-system of their discipline, and not to that of the facts being investigated. For 'relevance is not inherent in nature as such, it is the product of the selective and interpretative activity of human beings within nature or observing nature' (Schutz 1962: 5). The facts with which the natural sciences are concerned – molecules, atoms or electrons – mean nothing to themselves, and the interpretations of natural scientists mean just as little to them.

The facts with which the *social* scientist is concerned – the social world, human actions and their consequences – have, on the other hand, a particular meaning- and relevance-structure from the beginning. The agents 'have preselected and preinterpreted this world by a series of commonsense constructs of the reality of everyday life, and it is these thought objects which determine their conduct, define the goals of their actions, the means available for attaining them' (Schutz 1962: 6). Thus the social scientist has, in a 'theoretical attitude', to refer to 'thought objects constructed by the common-sense thought of man living his everyday life among his fellow-men' (Schutz 1962: 6). The theoretical constructs produced in research should not only supersede those of everyday life; they should also match them, if they are to be empirically valid. I may thus interpret Schutz's thesis as follows: the natural sciences are concerned with the formulation of constructs of the first order, which Schutz terms first degree. These constructs are those which do not have to relate to further meaning-structures. The social sciences are concerned with the formulation of constructs of the second order -'constructs of the constructs made by actors on the social scene [in the natural attitude], whose conduct the [social] scientist observes and tries to explain in accordance with the procedural rules of his [social] science' (Schutz 1962: 6). Consequently the most important task of social science methodologists must be 'the exploration of the general principles according to which human beings in everyday life organize their experiences, and especially those of the social world' (Schutz 1962: 59).

A social science wishing to achieve this goal must adopt the postulate of subjective interpretation, assume a subjective point of view. It must refer 'to the interpretation of the action and its settings in terms of the agent' (Schutz 1962: 34). The object of social research – human actions in all their different forms, their organization in various types of institutionalization and the products or artifacts produced by these actions – must be seen and analysed in and from the subjective perspective. For this a (re-)construction is needed which can structure processes of human action into certain types. It is only by doing this that the social sciences can investigate social reality in an empirically valid way. In other words, it is a question of understanding what agents mean when *they* act.

As already indicated, the meaning-contexts, the specific form of the stock of knowledge at hand and the relevance system of the agent in the everyday world, are all founded on the agent's unique, individual biographical situation. So how is it possible to grasp these subjective meaning-structures and reproduce them within the framework of an objective system of scientific theory, without the scientists becoming schizophrenic? How is this paradox to be resolved?

With the adoption of the theoretical attitude[20] an understanding of subjective meaning-structures becomes possible, so that the first aspect of the paradox can be eliminated. The formulation of constructs of the second order, which take appropriate account of the agent's subjective meaning, differs from the formulation of constructs of the first order in that the two constructs do not result from the same attitude. Therefore

> the thought objects constructed by the social sciences do not refer to the unique acts of unique individuals occurring within a unique situation. By particular methodological devices [. . .] the social scientist replaces the thought objects of common sense thought *relating* to unique events and occurrences.
>
> (Schutz 1962: 36)

To this end he develops models which represent the social world in the light of theoretically relevant problems.

The second aspect of the paradox, the question of the possibility in principle of reproducing subjective meaning in objective theories, is also closely connected with the difference between the 'natural attitude' and the 'theoretical attitude'. In contrast to the agent in the 'natural attitude', the researcher in the 'theoretical attitude' does not assume a 'here' in the social world, from which constructs might be formed. The 'corpus' of his or her discipline and its methods form the researcher's stock of knowledge. A researcher observes human action patterns and their consequences 'insofar as they are accessible to his observation and open to his interpretation. Their interaction patterns, however, he has to interpret in terms of their subjective meaning lest he abandon any hope of grasping "social reality"' (Schutz 1962: 40). The objective reproduction of subjective meaning-contexts is possible as long as the researcher does not share the same pragmatic relevance system as the agent under investigation. Thus this reproduction is only possible if the (social and physical) 'here' of the life-world is abandoned and a biographically determined stock of knowledge is replaced by the 'objective' knowledge of a particular discipline.

The theoretical approach to the everyday world

Since the reproduction, or reconstitution, of the everyday world takes place in the natural attitude, the theoretical attitude must relate appropriately to the former if it does not wish to bring on a crisis of alienated self-sufficiency.

All statements deriving from sociological theories should therefore be based on intersubjective experience of the natural attitude. From this, specific principles for sociological research can be derived.

In Schutz's view, the formation of sociological models and theories deriving from these general observations on cognition and theory should fulfil the following postulates:

1 *The postulate of logical consistency.* 'The system of typical constructs designed by the scientist has to be established with the highest degree of clarity and distinctness of the conceptual framework implied and must be fully compatible with the principles of formal logic' (Schutz 1962: 43). For it is their totally logical character which distinguishes scientific constructs (which replace everyday constructs) from everyday thought.

2 *The postulate of subjective interpretation.* 'In order to explain human actions the scientist has to ask what model of an individual mind can be constructed and what typical contents must be attributed to it in order to explain the observed facts as the result of the activity of such a mind in an understandable relation' (Schutz 1962: 43). This ensures that every action can be traced back to its subjective meaning.

3 *The postulate of adequacy.* 'Each term in a scientific model of human action must be constructed in such a way that a human act performed within the life-world by an individual actor in the way indicated by the typical construct would be understandable for the actor himself as well as for his fellow-men in terms of common-sense interpretation of everyday life' (Schutz 1962: 44). With the application of this postulate the alienation of social science from the everyday world can, in Schutz's view, be prevented.

Before considering how these general principles can be applied on the practical social research level, I shall first examine the most important fundamental assumptions of the phenomenological position.

THE CONSTITUENTS OF THE SUBJECTIVE PERSPECTIVE

The perspective of social research foreshadowed by Husserl, and elaborated in particular by Schutz, differs from Popper's 'objective' perspective in that it places the knowing subject's contribution to the meaning-content of the 'world' at the centre of discussion. All physical objects, ideal objects (language, etc.) and persons relevant to action are 'verified' *via* the subject's process of intentional constitution. In their analysis of social reality, social scientists must therefore take adequate and appropriate account of these commonsense constructs of the first order in their scientific constructs of the second order. The following are the constituent postulates for the subjective perspective:

1 *The postulate of relative realism.* Although there is a reality independent of the knowing subject, the latter can only perceive it *as* it appears to him or her, i.e. as he or she constitutes it, and not as it might exist 'as such' or 'in itself'. The starting point here is thus not the naive idea of a totally independent, objective reality but a relative meaning-structure of reality. This meaning-structure then becomes intersubjective if everyone carries out the same constituting processes, i.e. if reciprocity of perspectives makes possible an intersubjective experience of this reality. The constitution of meanings changes correlatively with the various attitudes.

2 *The postulate of intentionality.* Thought acts should not be seen as isolated wholes, but as the intentional consciousness-of-something which brings in both previous experiences of

knowledge at hand and the objects of present experience. A given fact should be understood neither as a purely physical nor a purely psychological one. It is intentional. It has just as much subjective meaning as it has physical qualities.

3 *The postulate of sociality.* The agent's stock of knowledge at hand is for the most part conveyed through social interaction ('socialization'), and is to a lesser extent the product of personal experiences in the life-world. Subjective acts of cognition always point to the intersubjective contemporary world and the intersubjective stock of knowledge. Reference to both aspects is made subjectively by the knowing subject. The socialized and therefore intersubjective knowledge available to the knowing subject forms the frame of reference for acts of explication.

4 *The postulate of subjective interpretation.* The agent's cognition process always takes place within the framework of the subject's relation to a stock of personal knowledge. Social reality should be understood as a meaningful construction by agents. Social scientists should therefore adopt, integrate and forefront the subjective perspective in their research.

5 *The postulate of empirical justification [Begründung].* The theoretical constructs of social research, which in the context of the subjective perspective are constructs of the second order, that is to say, of social research, should relate either directly or indirectly to verifiable facts in the everyday world. They should have intersubjective validity.

6 *The postulate of 'adequacy'.* The empirically based constructs of the second order, the products of the theoretical attitude (the researcher's acts of interpretation), should be compatible with experiences in the natural attitude. They should not alienate or be alienated from everyday experience. In addition, such theoretical constructions must satisfy the precepts of formal logic and contain a high degree of clarity explication and strict definition in the terms used (cf. above, p. 72).

MODES OF CONSTITUTION AND THE SITUATION OF ACTION

Like Popper, Schutz (1962: 74ff.; 1970: 73ff.) distinguishes – although he only does so implicitly – three worlds which an agent's natural attitude takes into consideration:

- the agent's subjective consciousness in the form of the stock of knowledge at hand (*subjective world*);
- the world of nature, of physical things as 'paramount reality' (the *physical world*), to which the agent relates the subjective world; and finally
- the *social world*, which comprises all other agents, their meaningful actions and the artifacts produced by them.

Although Schutz does not (1962; 1970) refer explicitly to it in his later works, his 'theory of life-forms' formulated in the 1920s does, in my interpretation, imply a relation of the different worlds to each other. The meaningful constitution of the 'ontological structure of the world' (Schutz 1970: 73) takes place for each ontologically different world according to the particular modes of constitution of the different 'life-forms'.

For every agent Schutz distinguishes six life-forms in which the ego has its existence. Each of them can be characterized as if it were a particular construct of experience and a corresponding system of the symbolizations of experience. They are the life-forms of 'pure duration', 'memory-endowed duration', 'the acting I', 'the Thou-oriented I', 'the speaking I' and 'the thinking I'. Each life-form is a system of meaning-establishment and becomes as Srubar (1981, 38) explains, 'more and more complex with increasing distance from the life-form of pure duration'. That is to say, each life-form has a corresponding level in which experience is symbolized, and every lower level is interpreted by the next highest.

My integration of the 'theory of life-forms' into an ontologically differentiated three-world model is the first attempt to interpret Schutz's claims from the perspective of his corpus overall. I shall now discuss it in more detail with respect to general social scientific methodology and the development of an action-oriented social geography.

Constitution of the subjective world

The subjective world is constituted in the 'life-form of pure duration'. In it the stream of experience flows on continuously and irreversibly in a constant 'becoming' [*werden*] and 'unbecoming' [*entwerden*], so that there is only ever a 'now'. More simply, it represents our non-attentive experiencing of the world

in all its fullness. Such experience cannot, however, be transformed in its entirety into the higher life-forms. The prepredicative experiences of the pure duration life-form are
selectively retained in the life-form of memory-endowed duration.
They become present-, past- and future-oriented. They are not an
undifferentiated 'now'. The aspects of consciously retained duration which are acted upon are those which are relevant to the
actual situation of the agent. The consciousness contains those
pure duration experiences which are capable of remembrance,
the agent's stock of knowledge. In fact pure duration experiences
only acquire meaning when they are remembered and ordered in
relation to a specific situation and according to the relevance
system in the natural attitude.[21] With his description of the life-
forms of pure and memory-endowed duration Schutz thus gives a
more complex version of the constitution of the stock of knowledge. For him the starting point is the pure duration experiences,
some of which enter memory-endowed duration – the stock of
knowledge – through selective symbolizations.

Constitution of the physical world

As we have already seen, in Schutz's view remembrance takes place
in relation to specific situation. The situation is influenced by the
location of the body [*Leib*], which is the centre of activities directed
towards the outer world. The function of the body, to quote Srubar
(1981: 32) again, 'is to mediate between duration and the homogeneous space–time world of extension. It converts memories into
spatial action'.[22] The physical location of the body, on the one
hand, structures the things which may affect the experience of the
'I' or self in pure duration, and on the other determines the direct
sphere of influence of the agent. The body [*Körper*] is thus the
'(particularly suitable) link' (Schutz 1982: 41) between the subjective and the physical worlds. As the vehicle and 'transit stage' of
cognition and action, the body determines the particular here and
now. Yet it does not itself determine the contents of experience.

The constitution of the physical world thus takes place through
the conscious self's experience of his own body in movement.
Although the meaning of such movement derives from conscious
intent, the consciousness as such is 'without the ability to carry out
an action' (Schutz 1982: 74). The body is therefore seen as the
functional link [*Funktionalzusammenhang*] between inner processes

and movements directed towards the outer world. On the one hand the body in the physical world becomes a medium of expression for the intentional consciousness, and on the other the spatial dimension is mediated and incorporated *via* the body [*Körper*]. Thus the physical or geographical location of the body affects the nature of pure duration experiences, and thereby this location affects memory-endowed duration.

What remain to be examined are the ways in which the physical world is constituted. The fact that the self experiences the body primarily in movement also means that it experiences even the body only *in*, and not *as*, a functional context.

> The experience of this movement is not transformed by the fact that I move [. . .]. Rather, the transformation takes place because this experience is necessarily [. . .] reinterpreted [. . .] as an experience of space. This opens up access to the world of extension; it alone mediates the experience *of* space and therewith of time and of external objects.
>
> (Schutz 1982: 88)

With the experience of the spatial character of one's own body, 'the *spatiality* of all other things has been discovered and given' (Schutz 1982: 104).

The constitution of the physical world thus remains bound up with the experiencing, acting 'I'. Because the 'I' or self draws the things of the external world into its own sphere of action, because it makes them the goal or means for subjective actions, they are experienced concretely through contact with the body. The moving and acting 'I' establishes that the physical world and the subject's own body have extension in common. Apart from the experience of the spatiality of the physical–material world, the subject experiences also the qualities of the various objects in relation to its own body, verifying them with correspondent meanings for its actions. The constitution of the physical world is thus bound up with the subject's body-mediated contact with those phenomena.

Among all the things that exist in the space–time world,

> a series distinguish themselves which are related to my body in still another manner [. . .] that is by extension in space and time. [. . .] These objects are similar to my body and thus awaken special attention. Most of all, they stand out because

they can be compared to my own past I. [. . .] In one word, they are *consociates.*

(Schutz 1982: 127)

Their actions, as expressive movements of their bodies, have their own meaning, based of course on duration and their stock of knowledge. The 'Thou' is constituted 'at the intersection of two durations, two memories and two courses of action: mine, of which I have primary knowledge, and those which I interpret as my experiences of him' (Schutz 1982: 127). We may conclude that the greater this correspondence of duration and memories is, the more likely it is that an intersubjective, similar constitution of the meanings of things, a reciprocity of meanings and their construction, will be achieved.

Constitution of the social world

Schutz (1982: 127) sees the starting level for the constitution of the social world in the 'life-form of the acting I in the Thou relation'. Things existing in the 'life-form of the acting I' in space and time cannot be known with sufficient certainty as long 'as I confirm its existence only out [of the perspective] of my own life and not also out of the course of the life of the Thou' (Schutz 1982: 128). That is to say, the intersubjectivity of the meaning of the social and physical–material worlds is only constituted in the 'life-form of the acting I in the Thou relation'. Such intersubjectivity is constituted mainly in the direct face-to-face situation. For here the agents' bodies face each other directly, and the reciprocal symbolizations and interpretations can be directly (reciprocally) checked.

The direct checking of symbolic gestures conveyed *via* the agent's body [*Leib*] is the key feature of the 'life-form of the acting I in the Thou relation'. Schutz distinguishes it from the 'life-form of the speaking I', the bipolar situation of talking. Here the symbolization of experiences in pure duration pass through a higher stage of complexity. Meanings are detached from the agent's body and re-presented in the communal, intersubjective meaning-context of the language system. 'Thanks to the conceptual-symbolic function of language, the experience is typified and generalized' (Schutz 1982: 143). From then on it belongs just as much to the 'I' as to the 'Thou', and will therefore be part of an 'objective' meaning-context. If the 'Thou' wishes nevertheless to grasp the subjective meaning of my words, it is necessary that his or her subjective access to the language coincides with my own access to this intersubjective meaning-context.

The 'life-form of the thinking I' [*Lebensform des begrifflichen Denkens*] is the precondition for the theoretical attitude to the everyday world. Srubar (1981: 52) comments that 'it is not clear whether by this life-form Schutz means scientific thinking only or also predicative thinking in the general terminology of everyday life'. I cannot go into this point here. To say that this life-form is a precondition for the theoretical attitude must suffice. In the most general sense this life-form can be described as a continuation of the level of the speaking self's intersubjective symbolizations. It thereby represents the sphere furthest away from subjective experience of inner duration. Its highest level is formal logic and 'objective' knowledge is its goal. This continuation must, however, be adequately carried out: it has its foundations in each of the lower life-forms, from the thinking self down to pure duration.

If this continuity is not present, any formalization is beside the point. That is to say, the results obtained in the theoretical attitude in the 'life-form of the thinking I' must adhere to the postulate of adequacy and supersession, regardless of whether these results refer to the subjective natural attitude and its constructions, and regardless of whether they refer to the physical or social world.

Having examined the 'theory of life-forms', we can now characterize the three world spheres in detail. The *subjective world* is concerned with a stock of knowledge acquired by the I and transmitted socially. This stock of knowledge is biographical in nature, and is lifted out of the life-form of pure duration by means of the life-form of memory-endowed duration. As an entity this background knowledge remains unproblematical. Only that part which agents use and classify for their interpretations of the action situation is put to the test.[23]

This subjective world of the natural attitude relates to the physical and social worlds. The subject constitutes the physical world, the world of physical objects, via the experience of his or her own body in movement, relating to it through the stock of knowledge. This is a sphere where there are no expectations which could be discussed – the agent has to adapt to that sphere's ontological constraints.

As I said, the *social world* is experienced by the agent primarily through the 'Thou', especially in the form of other agents' typical expectations in typical situations. An agent has to communicate with other subjects regarding the meaning-content of subjective actions, so that successful interactions may take place. Thus in action situations the ontologically different worlds are brought into relation in the meaning horizon of the subjective world (see Figure 5).

The agent accordingly takes in more and more world relationships in the action situation, with every *definition of an action-situation*, directly or indirectly, that agent always refers to the 'defined reality of physical things'. In the definition of the action situation, an order (of meaning) is established by the subject, into which the different world relationships are integrated. Definitions of the same situation by several agents will *a priori* deviate from each other. By means of a series of idealizations, these deviations in the natural attitude of everyday experience are as a rule ignored, unless there is obvious dissent, or unless interaction with others becomes problematical. Should interpretations of a situation diverge so greatly from one other that intersubjective communication is no longer possible, the participants must change and expand the type-structure of their stock of knowledge. This must be done in such a way that reciprocity of perspectives is re-established, and the problematic province of the life-world is once more part of the unquestioned everyday world, and intersubjective communication and coordination of action again become possible.[24]

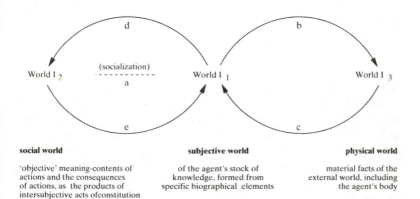

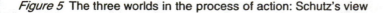

Figure 5 The three worlds in the process of action: Schutz's view

EMPIRICAL SOCIAL RESEARCH

Alfred Schutz defines the primary objective of every scientific action as 'clarifying the confused judgements of daily life through investigation and elucidation' (Schutz 1974: 339) whether this is carried out within the framework of natural science or social science. Although research in natural and social science have common objectives, Schutz sees differences especially in their empirical approach.

Max Weber's (1968: 1) famous formulation of the general objective of social research as the 'interpretative understanding of social action and thereby the explanation of its course and consequences' is taken up by Schutz. For him, Weber's formulation means that the aim of empirical social research should be: (i) to understand everyday actions within the framework of their reasons and goals; (ii) to grasp empirically adequately the subjective constructions of the social world; (iii) to reproduce them in objective, general and social scientific constructions. With such constructions the social scientist should eventually be in a position to interpret the social world adequately and to understand it so that he does not alienate himself from the agents' 'subjective meaning'. Schutz concentrates on this goal, but he incorporates the results of the phenomenological epistemology, in order to remove the central paradoxes of Weber's methodology.

To clarify further this idea of social research, Schutz proceeds *via* a comparison with the postulates of the methodologists Hempel and Nagel.[25] He agrees with them that any form of empirical research entails processes of controlled deduction. He also agrees that general precepts should be formulated which should be subject to empirical testing, and 'that "theory" means in all empirical sciences to explicit formulation of determinate relations between a set of variables in terms of which a fairly extensive class of empirically ascertainable regularities can be explained' (Schutz 1962: 51–2). Theory must satisfy the ideals of the thematic unity, simplicity, universality and precision. Schutz also agrees with Hempel and Nagel that the narrowly restricted universality of the theoretical propositions of sociology does not 'constitute a basic difference between the social and the natural sciences, since many branches of the latter show the same features' (Schutz 1962: 52).

Schutz does not, however, share Hempel and Nagel's view of the testing stage of theoretical statements. Nor does he share their interpretation of Weber's postulate of 'the interpretation of subjective meaning'. Schutz argues that these methodologists fundamentally misunderstood Weber's postulate because they did not interpret its basic point correctly. They 'identify experience with sensory observation, and [. . .] assume that the only alternative to controllable and, therefore, objective sensory observation is that of subjective and, therefore, uncontrollable and unverifiable introspection' (Schutz 1962: 52). Schutz rejects the process of introspection in the same way as he rejects mere sense observation. 'Subjective understanding' [*verstehen*] for him does not mean 'that [the] understanding [. . .] of another man [. . .] depends upon the private, uncontrollable and unverifiable intuition of the observer, or refers to his private value system' (Schutz 1962: 56). 'Subjective' can in no way refer to the researcher. 'Subjective understanding' refers rather to the actions to be investigated, which must be understood within the framework of their reasons and goals. Similarly, the products of human acts, the artifacts, should be investigated 'in terms of the purpose for which it was designed by [. . .] a fellow-man and its possible use for others' (Schutz 1962: 56). 'Subjective understanding' is therefore concerned to discover 'what the agent [him or herself] "means" by his [or her] action, in contrast to the meaning which *this action* has for a neutral observer' (Schutz 1962: 57). Schutz's most important methodological postulates are based on this point of departure: the 'subjective interpretation' and 'adequacy' of scientific constructions in relation to those of everyday life.

Both postulates are related to each other: subjective interpretation should satisfy the postulate of adequacy, and that can only happen if the subjective interpretation is given. Therefore we must first turn our attention to subjective understanding, to subjective interpretation as a cognition process of agents in the natural attitude, and as a sociological cognition process in the theoretical attitude. Similarly, we must look at the 'problem of adequacy', both within everyday experience and in terms of the relationship between science and the life-world.

Subjective interpretation

For Schutz, 'understanding' in the everyday natural attitude means interpreting our own actions and those of the others with whom we interact. However, we are not limited to the position of observers who only register movements through sense perception. We have rather a stock of knowledge at hand which contains categories which allow us to grasp what the other 'means' by his or her actions from our own standpoint, we can do this without having to resort to self-conscious introspection.

If social scientists wish to do justice to their object of research, they must use the same *method* in the theoretical attitude as agents use in the natural attitude. But they must do so with 'explicit clarity'. The postulate of subjective interpretation thus means that social explanations should be concerned with the subjective meaning of action, and hence with the subjective constitutions of social reality. From the methodological point of view, the differentiation between subjective and objective meaning, however, deserves further investigation.

Schutz (1972: 29ff.) uses the following example to illustrate the difference between subjective and objective meaning. Let M be the meaning which an action A has for the agent X. This bodily action is perceived by a 'naive' observer NO in the natural attitude, and by a scientific observer SO in the theoretical attitude, and understood by both NO and SO. M denotes the intended meaning of X, and M' and M" the meaning-content of A which NO and SO attach to the action A. What does this mean? First, that NO and SO attach a specific meaning to the outer course of the action A, but that this meaning cannot be absolutely identical with the intended meaning M which X gives to his action. Drawing on the personal stock of knowledge NO will attach the meaning M' to the action, and SO, drawing on his sociological scheme of interpretation, the meaning M".

NO and SO thus interpret action A in such a way that the process A acquires 'meaning' within *their* interpretational framework. They will have the same interpretation of A if their stock of knowledge with regard to it is the same, and they will intersubjectively attach the same meaning to that process. This endowment of meaning refers back to 'some meaning-endowing act [*Sinngebung*] of yours with respect to the same world. Meaning is thus constituted as an intersubjective phenomenon' (Schutz 1972:

32). If this precondition is given, we can refer to M′ and M″ as the 'objective meaning' of A. Thus if a social phenomenon is to acquire an objective, i.e intersubjective, meaning, the most important precondition is a reciprocity of perspectives of intended meaning-constitutions.

From a purely methodological point of view, consider Husserl's statement that 'each meaning-formation [*Sinngebilde*] can be asked about its *essentially necessary meaning history* [*Sinnschicht*]' (Husserl 1969: 208). Two kinds of analysis follow for this: one 'static', the other 'genetic'. In *'static analysis'* meaning-formation can be examined apart from its *genesis*: account is no longer taken of the intentional acts of its producer. The phenomena of the outer world 'which present themselves to me as indications of the consciousness of other people' (Schutz 1972: 37) are then observed and interpreted *in themselves*.[26] We turn our attention to phenomena in the objective sense. The interpretation of the objective meaning of a phenomenon 'is exhausted in the ordering of the interpreter's experiences of the product within the total meaning-context of the interpretative act' (Schutz 1972: 134). So at the same time the interpreter leaves out of consideration those personal meaning-contents which the product has for the producer. 'Objective meaning therefore consists only in a meaning-context within the mind of the interpreter' (Schutz 1972: 134), 'abstracted [. . .] separated from the constituting processes of a meaning-endowing consciousness, be this one's own or another's' (Schutz 1972: 37). With this kind of 'turning-towards' [*Zuwendung*], 'only' the general meaning of a phenomenon is grasped, that which concerns the intersubjective ('objective') meaning-context, whether it is that of a 'naive' observer or that of a social researcher operating within the discipline's theoretical framework.

'Genetic analysis' means that an event in the social world is seen as the sign of an intentional experience of an agent's consciousness. For the agent this sign indicates a 'subjective meaning'. In this mode of 'turning-towards' [*Zuwendung*], interpreter NO or SO is no longer concerned with the event 'in itself' but with the consciousness of the other person and the other's constituting process. The event A presents an indication of their consciousness. We are now turning towards social phenomena in their 'subjective meaning'. This is when we 'have in view meaning-context within which the product stands or stood in the mind of the producer'

(Schutz 1972: 133). Discussion of the subjective meaning of the social world has as its object an understanding of the constituting processes of the person 'who produced that which is objectively meaningful. We are therefore referring to his "intended meaning", whether he himself is aware of these constituting process or not' (Schutz 1972: 37).

The question now, however, is with what aspect of the agent's constituting processes 'understanding' should concern itself: the constitution of the stock of knowledge, or the constitution of meaning? Methodologically we must clarify whether either of the two aspects should be made paramount in the subjective interpretation and explanation of human actions.

In discussing these constitutions, I pointed out that the constitution of the stock of knowledge acquired up to a given biographical point (first aspect) was always the precondition for any intentional meaning-constitutions (second aspect). The possible intentions and purposes of action are the 'in-order-to' motives. In the action context, they are always limited by the 'because' motives, which stem from the stock of knowledge at hand.[27] The possible intentions of actions are therefore limited by the biographical stock of knowledge.

This problem can be explained through the use of the following example. Someone kills someone else and robs him of his money. How can we explain it?[28] If we go for the 'in-order-to' motive – the purpose – the explanation could be: x killed y *in order to* get his money. If we go for the 'because' motive, the explanation might be: x killed y *because* he grew up in a criminal milieu. In other words: does the understanding of an action involve reference to the origin of the intention, or to the intention itself as the aspect which directs the action?[29]

In part, for Schutz (1972: 93), the subjective meaning at the time of the action is constituted in the project, in the purpose, the 'in-order-to' motive. The genuine 'because' motives go further back, but are not present in the agent's field of vision at the time of the action. Thus if we wish to grasp the subjective meaning of an action, we must first concentrate on the 'in-order-to' motive. Later we have to, as can the agents themselves, turn our attention to the subjective preconditions of the action, which form the range of possibilities of agents' projects.

If we were to consider 'because' motives only, however, we would be attempting to give a causal or structural explanation of actions. We would be trying to answer the question 'why?'. But since 'because' motives do not play a direct part in the agent's considerations at the time of the action, Schutz sees every motivated action as a (not causally) determined reference to the agent's stock of knowledge, and, at the same time, as a spontaneous act. It follows that sociological research applying the postulate of subjective interpretation in its explanation of human activity must concentrate primarily on 'in-order-to' motives, and not only on the 'because' motives. The latter will become important for understanding of origin of subjective meaning.

Adequacy

Agents understand another agent in the *natural attitude* if they can adequately interpret the meaning which a particular action has for the other. In order to do this they try to understand the course of a co-agent's experience by concentrating on his or her 'in-order-to' motive and, in certain cases, also on the 'because' motive. Agents can, however, only approximately understand the other's purpose by means of their own schemes of interpretation – they cannot use that of the co-agent. Thus interpreters project the goal of the other's action in terms of their own action goals and elucidate the other's motives accordingly. For this they need certain idealized 'types' of 'in-order-to' and 'because' motives, so that they can gain access to the meaning-context of their contemporaries' actions, and be able to grasp their modes of constitution.[30]

With the postulate of adequacy Schutz is making two stipulations. First, the procedures of the social sciences should be compatible with those applied in everyday actions. Second, the knowledge acquired in this way should be empirically valid. In this view the first requirement is met by the formation of ideal-type models which correspond to the everyday mode of communication in the contemporary[31] situation of action.

Schutz (1972: 185) regards the model-based scheme of interpretation as a kind of 'empty formula' [*Leerformel*],[32] under which an agent attempts to subsume the empirical meaning of the other's action, in the process of understanding [*Verstehensprozess*]. If it succeeds, the ideal type has for the time being been confirmed, and adequate understanding of the meaning has taken place.

These models may refer to two things: first to the self of the other person, who generates certain products which have to be interpreted; and, second, to the

> expressive process itself, or even of the outward results which we interpreted as the signs of the expressive process [completed actions and artifacts]. Let us call the first 'personal ideal type' [*personaler Idealtypus*] and the second the 'material' or 'course-of-action type' [*materialer Idealtypus oder Ablauftypus*].
>
> (Schutz 1972: 187)

In constructing personal models we postulate the other's consciousness, hypothesizing its stock of knowledge as invariant. We thus project the model of an agent whose particular course of action, and particular product, might in typical mode be placed in a particular subjective meaning-context. In constructing material models (course-of-action types), we refer to previous similar repeated courses of action, i.e. completed actions involving similar means to achieve the same goals (showing similar 'in-order-to' motives) and arising from the same genuine 'because' contexts. The generally typical motive of the course of action is seen as invariant.

The two kinds of model construction are closely related. 'The personal ideal type is therefore derivative, and the course-of-action type can be considered quite independently as a purely objective context of meaning' (Schutz 1972: 188). For instance, as an agent in everyday life I may try to construct the personal type 'carpenter'. To do this I must first have in mind a course-of-action type relating to his action (as an objective context of meaning). For I can only construct the type 'carpenter' on the basis of the typical action of people who have this job, and not by referring to the specific features (father of two children, etc.) of the performer of the job: the personal type presupposes the prior existence of a hypothetical course-of-action type. The course-of-action type *may* also be traced back to a personal type, and this is frequently necessary. Such a transition takes place when we 'imagine the corresponding subjective meaning-contexts which would be in [the agent's] mind, the subjective contexts that would have to be adequate to the objective contexts already defined' (Schutz 1972: 187–8). As noted above, the subjective contexts are the genetic aspects, the objective contexts the static aspects. The transition thus takes place from the static aspect to the genetic aspect. This procedure is necessary when the adequacy of the assumed objective meaning-context is called into question, or if it has to be verified.

Thus a two-fold technique of ideal-type comprehension is possible in everyday situations of understanding. On the one hand, we can attain understanding by starting with a typical product (act/artifact) and inquiring, *via* a typical action, into the personal type. On the other, we can understand the type of the agent *via* his or her action, and enquire from there into the typical product.

Ideal types or models constructed in the natural attitude show varying degrees of anonymity. This depends on the closeness or distance of the 'contemporary agent acting in the everyday world' to the matter to be typified. If the stock of knowledge with regard to a type is highly complex, it will be rich in specific content. On the other hand, the greater the anonymity of the ideal-typical construction, the greater will be its span of validity or application. 'We can say, then, that the concreteness of the ideal type is inversely proportional to the level of generality of the past experiences out of which it is constructed' (Schutz 1972: 195). In other words, the more abstract, general and anonymous the construction of an ideal type, the more universal the span of validity of the ideal-typical schema.[33] Schutz accordingly (re)constructs the following stages of ideal types in order of increasing anonymity: they move from 'characterological' types to 'habitual' types, through to the personal ideal types of social collectives (i.e. of agents participating in them) such as the press, the economy, the state, etc., then on to the ideal types of closed sign systems (e.g. different languages), and culminate in the ideal types of cultural objects, ideal types of artifacts, which have the greatest degree of anonymity.[34]

As to the verification of the adequacy of models, Schutz points out that any type/model can only ever be hypothetical: there is always an element of chance present. Communication will take place adequately in an action situation in the contemporary world if there is sufficient congruence between the ideal-typical interpretative scheme and the meaning-content to be grasped. All attempts at communication in the natural attitude are made by means of 'ideal-typical constructions, and the invariant motives contained in them, which are hypothetically assumed to be mutual' (Schutz 1974: 285).

In the *theoretical attitude*, social scientists should do their utmost to ensure the continuity of the construction of everyday types. That is to say, the transformation of everyday ideal types into scientific ideal types should be characterized by controlled idealization and formalization. The postulates of subjective interpretation and adequacy thereby acquire additional content in the theoretical attitude. If they have to be fulfilled in the natural attitude so that agents can communicate in the everyday world, they now become the precondition for compatibility between constructions of the first and second degree. This is to prevent the everyday experiences of a researcher from being inadvertently removed to the theoretical level (cf. Schutz 1972: 190).

Although social scientists have exactly the same access to the social world as the ordinary everyday observer, they differ from the latter in terms of the experience-context on which intersubjective understanding is based. This experience-context – or rather, relevance system – should be determined by the overall objectives of the discipline. Social scientists' relevance system should never, in contrast to the everyday relevance system, refer back to their naive day-to-day experience but should be based only on explicit positional acts of judgement, on constituted ideal objectifications, that is to say, on conclusions of thought, 'and never on prepredicative acts of laying hold on [*in Selbsthabe erlebte Erfassungen*] the other person himself' (Schutz 1972: 223). The researcher's relevance system should build on the empirically based stock of knowledge of the discipline, on its existing theoretical framework.[35]

As with everyday procedure, social scientists should also adopt a two-fold technique of type construction and its application. The construction of models of the social world, with which the social scientist must replace meaning-constructions of the first degree, should start out 'to construct typical course-of-action patterns corresponding to the observed events' (Schutz 1962: 40). These typical patterns should refer to a personal type: the model of an agent and an agent's consciousness, but it cannot be as comprehensive as that of an actual agent. It is limited, and has at its disposal only those elements of the action pattern which are relevant for the problem in hand – the problem formulated by the scientist in the 'theoretical attitude'. Building on this, the social scientist constructs typical patterns of action or courses of action, assigning them to the model of the agent. Schutz (1962: 41, 64) calls this model a 'homunculus' or a kind of marionette.

The personal model thus comprises a limited structure of invariant goals of action: 'A merely specious consciousness is imputed to them [. . .] in such a way that its presupposed stock of knowledge [. . .] would make actions originating from it *subjectively* understandable provided that this action were performed by *real* actors' (Schutz 1962: 41). Schutz is here merely stating a hypothetical condition which enables the postulate of adequacy to be taken into account. The homunculus, however, is not subject to the same ontological conditions as the actual agent, for he is after all a construct developed in the 'theoretical attitude'. The homunculus can only choose between alternatives placed before him by the social scientist. 'He cannot err if to err is not his typical destiny. [. . .] He is nothing more than the originator of his typical function' (Schutz 1962: 41).

The social scientist constructs the stage of the social world and 'distributes the roles, who gives the cues, who defines when an "action" starts and when it ends and who can determine, thus, the "span of projects" involved' (Schutz 1962: 42). This means at the same time that the typifying sociological experience is also 'an objective meaning-context whose object, however, is subjective meaning-context (to be precise, the typical subjective process of personal ideal types)' (Schutz 1972: 224).

Now Schutz's model construction could easily sound like the economists' model of 'homo economicus'. In order to counter this misconception we must distinguish clearly between two things: the rational construction of scientific models of human action, and the construction of scientific models of rational action.

The rational construction of scientific models has already occurred if the postulate of logical consistency is followed. Such a model cannot, however, relate to non-rational types of action. In contemporary economic and location theory research, ideal types of rational and optimal action are usually projected. With empirical problems, however (an economic report or something similar), it is usually assumed that the actual agents act or have to act just as the ideal type constructed by the scientist would act. It is precisely this misconception, which leads to such an identification, that Schutz seeks to avoid in his proposals for sociological research practice. He demands that sociological ideal types should 'not only be compatible with the established conclusions of all the sciences but must explain in terms of motivations the very subjective experiences which they cover' (Schutz 1972: 224). To put the point in Max Weber's terminology, these postulates of Schutz mean that the ideal types of human action constructed by the social sciences must possess at the same time both 'causal-adequacy' [*Kausala-däquanz*] and 'meaning-adequacy' [*Sinnadäquanz*].[36]

With regard to the *criterion of causal adequacy*, Schutz (1972: 231) follows Weber (1951: 286ff.) in emphasizing that 'when we formulate judgements of causal-adequacy in the social sciences, what we are really talking about is not causal necessity in the strict sense but the so-called "causality of freedom", which pertains to the end–means relation'. A type construct is causally adequate, then, if it is probable 'that, according to the rules of experience, an act will be performed (it does not matter by whom or in what context of meaning) in a manner corresponding to the construct' (Schutz 1972: 232). The requirement here is that there must be at least one empirically established action which corresponds to this construct. The construction of models must therefore fulfil two further conditions in the theoretical attitude. First, the construction of types must refer to actions which are repeatable, i.e. it should have at least one precedent. Second, in 'a type construct of ordinary purposive action the means must be, in the light of our past experience, appropriate to the goal' (Schutz 1972: 233). The requirement of causal-adequacy means, then, that the motives regarded as invariant in the ideal type were indeed effective during the course of the action, or: to say that the motives must be causally adequate means merely that the motives could have brought about this action and, more strictly, that they probably did so' (Schutz 1972: 236). We have now addressed to some extent the question of how far causal-adequacy is bound up with meaning-adequacy.

Schutz (1972: 233) posits the thesis that 'all causal-adequacy which pertains to human action is based on principles of meaning-adequacy of some kind or other'.[37] Causal-adequacy means 'the consistency of the type construct of a human action' – the invariant assigning of particular motives by the observer – 'with the total context of our past experience. Furthermore, we can come to know a human action only by ordering it within a meaning-context. [. . .] Causal-adequacy, then, insofar as it is a concept applying to human conduct, is only a special case of meaning-adequacy' (Schutz 1972: 233–4). In contrast to Weber, Schutz emphasizes that when establishing meaning-adequacy we should refer not to the objective meaning-context presupposed by the observer, but to the agent's subjective meaning-context. The subjective meaning-context which is projected as invariant in the model must also be valid for the agent. That is to say, our construction should not contradict other experiences we have had regarding the agent but agree with the personal model which is valid for him or her. This first condition establishes that there should be compatibility between the course-of-action model and the personal model, whereby the social scientist 'would have a completely free hand in his construction of a personal ideal type' (Schutz 1972, 235). This should not happen: the personal model and the course-of-action model must always be tested in relation to actual agents and actions.

SUMMARY

In phenomenology, meaning in the social world is intentionally constituted on the basis of the stock of knowledge. However, in spite of the key role of the subject in the process of knowledge an 'objective' social world still exists. This is explained by the fact that, on the one hand, the stock of knowledge is the product of social-ization, and, on the other, by the intersubjectively valid con-stitution of meanings which build on that stock of knowledge, and which are assumed 'for the time being', to be true. Pheno-menologists contend that subjects' intentional acts are carried out in the everyday world within the framework of the natural attitude.

In Schutz's view, scientific research into the social world in the theoretical attitude is only relevant if it takes into account the postulates of logical consistency, of subjective interpretation and of adequacy. His basic idea is that the researcher in the theoretical attitude must set out to understand adequately the subjective mean-ings which agents intentionally constitute in the natural attitude. That researcher must also develop (personal and material) models which satisfy the criteria of meaning-adequacy and causal-adequacy.

Finally, I have pointed out that Schutz was explicitly opposed to any approach which advocated sociological research based on behaviourism. Since behaviourist research denies or at least ignores intentionality in human actions, it is fundamentally in-compatible with phenomenological epistemology. The pheno-menological epistemology of social science, no more than Popper's objective perspective, does not justify any behaviourist or quasi-behaviourist approach. On the other hand, as we shall see, phenomenology is a crucial point of departure for a new approach to social geography based on action theory.

Chapter 4

An epistemological synthesis

In this chapter I compare the critical–rational and pheno-menological perspectives on social research. I shall reopen the question of whether the two positions are mutually exclusive. Theoretical geographical discussion[1] and sociological debate[2] assume that they are. As I said in Chapter 1, I think there are also grounds for regarding them as complementary.

We have seen that Popper and Schutz start out from dia-metrically opposed theses regarding the theory of knowledge. Popper is concerned to found a theory of knowledge without a knowing subject. Schutz, on the other hand, following Husserl, postulates a subjective theory of knowledge in which an active role is assigned to the knowing subject. For Popper, both world 1 (physical world) and world 3 (objective ideas in themselves) exist independently of the knowing subject. They are, although onto-logically different, objective in character. No matter whether it is a person acting in the everyday world or a scientist, cognition pro-cesses are deductive. On a world 3 basis, the agent draws con-clusions deductively about the physical–material and social circumstances of the action situation. The tasks of science are critical enlightenment, and narrowing the gap between our knowl-edge and 'objective truth in itself'. Common sense and scientific actions are not distinguished by their logical structure, but by the degree of 'enlightenment', and the closeness to truth, underlying the actions performed.

Schutz also, implicitly at least, starts from the idea that there are three ontologically different worlds. For him, too, the world of nature and the social world exist independently of the ego. They can only achieve meaning, however, through subjective meaning-constitution based on the stock of knowledge. This applies both to the meanings of material things in the physical world, and to the circumstances of the social world. For Schutz the two worlds differ from each other in that the first does not have its own meaning- and relevance-structure. The natural world and the social world are constituted in their meanings by the subject acting in the everyday world in the 'natural attitude'. The scientist or researcher, operating in the 'theoretical attitude', must take ade- quate account of these meanings. The agent in the action situation draws conclusions about various facts and circumstances accordingly.

While Popper and Schutz both start out from the idea of three ontologically different worlds, the difference lies in the fact that for Schutz, the subjective meanings of the various elements of the different worlds are all-important. Popper, on the other hand, adheres to the objective perspective and reconstructs contexts without referring initially to the subjective world: he sees the latter as consisting primarily of world 3 elements.

Where the critical–rational and phenomenological approaches diverge most widely is over the realism postulate, and conclusions based on it. Popper's realism postulate is absolute. He starts from the premise that both the objectively pre-given physical world and the objectively existing world 3 represent conditions for verifi- cation, or falsification, of circumstances and facts in the mental world. That area of world 3 which Popper calls the social world (above all, social institutions and traditions) achieves objective status through the fact that it arose out of the *unintended* by- products of intentional human actions. In addition, the facts and circumstances of the social world are just as real as those of the physical world, for both can become important factors con- straining human action. Ideas 'as such' also have objective status until they are conclusively refuted.

Schutz's realism postulate is a relative one. All reality exists for the knowing self only in the form of its meaning as it appears to the subject, and not as it might exist as such. Husserl and Schutz do not deny the existence of 'objective' facts, but place them in a different context. They do not assign to them the status of an existence independent of the agent, but see them as the product of the subject's acts of cognition, as facts whose meanings are intersubjectively constituted. In other words 'objective meaning' signifies for Schutz something akin to the 'generally valid mean- ing' of a fact or circumstance, to an intersubjectively shared meaning-context. With each act, the agent is checking whether subjective meaning-endowments are shared by others: 'the stock of knowledge is put to the test'.

This checking procedure differs from Popper's falsification of subjective beliefs through action in that if the result of the test is negative, the knowledge has not conclusively been proved false. The agent may learn something about the partner in interaction, and possibly achieve better communication if their points of divergence are removed. But the partner cannot be regarded as a knowledge falsification test, because their intersubjective communication cannot exist independently of reciprocal interpretations.

On the other hand, it becomes apparent in this context that Schutz also, in Popper's terminology, presupposes a deductive principle on the part of the agent, in both common sense and scientific actions. In neither do agents proceed according to inductive means. The subject acting in the everyday world puts a stock of knowledge acquired there to the test, and the researcher, observing agents in the everyday world, does the same with 'all the axioms, fundamental principles, theorems, and deductions of a discipline' (Schutz 1972, 223).

Critical rationalists start from the basic idea that there is a reality independent of the agent, which can be used at any time to test the validity of theoretical statements. Yet as we have already seen, this test requires some form of mediation. A relation must accordingly be set up between subject and object by means of sense data guided by theory or hypothesis. The empirical testing of theories on reality independent of the agent should be carried out mainly by means of sensory perception. Observations and observational situations should be controlled in such a way that they may be tested objectively at any time.

Popper sees experience as an active process based on hypotheses: only the latter can 'illuminate' reality. Since hypotheses always take the form of linguistic statements, observations can only be made through the medium of language, i.e. they consist of elements from world 3. In order to establish whether a theory corresponds to reality, i.e. is empirically valid, there must first be agreement regarding the meaning-content of the (descriptive) terms contained in a hypothesis (on which the observation is based). It is then possible to check whether all the participants are referring to the same object of reality, and whether their sense perceptions are directed towards the same object, i.e. the same features of that object. This problem can be resolved through the procedures of explication and definition. In order to determine whether there is congruence between the (theoretical) statement and reality, it is also necessary to have a truth criterion. Popper sees this as the degree of *correspondence* between a theoretical statement and the fact it refers to. This truth criterion also involves the problem of the operationalization of (descriptive) terms, the transformation of conceptual meanings into indicators which can be experienced through the senses. For the precise verification of theories exact measurements are also necessary, which must comply with the principles of the correspondence theory of truth.

Schutz's arguments evidently point to sociality. In contrast to Popper, Schutz starts from the idea that it is not the biological presuppositions of the knowing subject (biologism postulate) which are the most important component of the knowledge available, but the socialized elements of knowledge. The agent's contemporary world is only 'real' when it has become a 'social contemporary world [*soziale Mitwelt*]'. 'Reality' thus varies according to the nature of the subject's stock of knowledge and his corresponding relevance systems. As Berger and Luckmann (1966: 15) have pointed out, 'what is "real" for a Tibetan monk may not be "real" to an American businessman'. And here we come to a further important aspect of the phenomenological understanding of empirical reality: if what is constituted as 'real' depends on the agent's stock of knowledge, it cannot be assumed that there is an independent (universal) reality which can be used for objective verification. The only such reality is the social reality which is limited in space and time. The empirical sphere is thus characterized by a social, temporal, spatial and factual relativity: the fact of social relativity is expressed in the relationship between 'knowledge' and 'reality'. 'Specific agglomerations of "reality" and "knowledge" [obviously belong] to specific social contexts, and these relationships [. . .] have to be included in an adequate sociological analysis of these contexts' (Berger and Luckmann 1966: 15).

The empirical testing of theoretical statements and assertions from the subjective perspective can be illustrated by the following example. A tree need not, in the 'natural attitude', be interpreted as an objectively natural object; it may be real as the dwelling-place of good or evil spirits. In the appropriate social context it is not necessary to question or disprove this interpretation. If the interpretation 'dwelling-place of good or evil spirits' is inter-subjectively confirmed in the everyday world, it 'corresponds to' the social phenomenon and can thus count as a component of social reality. The connotation of 'real' may later be lost if the interpretation 'dwelling-place of spirits' is refuted or qualified by other experiences of reality. A researcher's statements about social reality are only seen as valid if *the meaning* of the tree in the respective social context has been grasped.

In the phenomenologists' view, then, the criterion of validity of empirical statements about reality is meaning-adequacy. The goal of empirical social research is the adequate comprehension of the relevant meaning-contexts. The findings of empirical research are 'true' if they satisfy the criterion of 'meaning-adequacy'. This idea of the 'truth concept' should not, however, be misconstrued. Phenomenologists do not believe that in contrast to scientific or theoretical supposition, everyday experience and truth in the con-temporary world are never mistaken. Although they perceive 'truth' as specific to the situation and the subject's constitution of it, they do not see 'everyday truths' as infallible. Their goal is the objective representation of subjective meaning. Neither do pheno-menologists wish to equate the scientist's or researcher's subjective opinions with 'objective truth'. In Schutz's account of the 'theore-tical attitude' he expressly points out that the scientist should 'bracket' pragmatic interests, replacing them with the relevance system of the discipline.

We can therefore conclude that Schutz's 'objectivity' require-ment and Popper's 'value-free' postulate are identical in their meaning. As far as the social sciences are concerned, Popper and Schutz both postulate the existence of objective facts, but place them in different contexts. It is my contention that Schutz's inter-pretation of 'objective facts', that they have intersubjectively accep-ted meaning, does not contradict Popper's arguments. Popper accepts uncritically the existence of 'objective social facts', and Schutz enquires into the conditions under which they originate and continue to exist. Schutz is not contradicting Popper when the latter goes on to assert that certain forms of society enable human beings to survive better than others.

Popper does, however, point out that the objectivity of social facts and problems is based on the fact that they are the un-intended consequences of actions, and that new problems con-tinue to arise from the interplay of the unintended by-products of such actions. Bearing in mind that Popper stresses this particularly in order to prove the error of the 'conspiracy theory' and to reject the subjective perspective, we can, without ourselves succumbing to the 'conspiracy theory', put the following opposing argument. Proof of the independence of the negative consequences of agents' intentions and their acts does not imply a rejection of the subjective perspective. The point is precisely that only the intended consequences are subsumed under intention, and a causal-adequacy relationship is reconstructed between intention and consequence. If these requirements are met, as they are by Weber and Schutz, Popper's fears appear to be unfounded. Yet, in turn, this does not mean that Popper's assertion that most social problems arise from the unintended consequences of actions should be rejected. We should only be succumbing to the 'conspiracy theory' if in spite of empirical proof to the contrary an unintended consequence of actions were to be identified as an intended one.

In fact Schutz sees the relationship between 'realistic-objective' approaches in social science and the subjective approach as complementary. For him the postulate of subjective interpretation simply means not only that we *can* point to the actions in the social world and their subjective interpretations, but sometimes we *must*. In the context of Popper's arguments, we can interpret this postulate as follows. In certain circumstances the objective consequences of actions can be investigated as such, but in others it is necessary to apply the postulate of subjective interpretation. If social scientists wish to repair 'objective' problem situations, then according to the interpretation of Schutz's arguments presented here, they must first establish the meaning-contexts of the actions which led to negative consequences. Only then can they persuade the perpetrators of those actions to adopt less problematical ones. If such meaning-contexts are generally known, if the actions took place in an intersubjectively recognized meaning-context, we should not, in research, have to adopt the subjective perspective.

Popper sees the goal of the social sciences as the investigation of the objective consequences of actions and not of the origin of the actions themselves. We have seen that Schutz indicates that he also finds this procedure meaningful in certain circumstances. His 'static analysis' investigates social facts separately from their initial endowment of meaning: the social scientist enquires into phenomena in the objective sense. The consequences of actions are analysed and interpreted within the framework of the 'theoretical attitude' of the respective discipline. This procedure thus shares important features of Popper's 'biological view' of the social world. At the same time, Schutz also sees static analysis as a preliminary stage to the understanding of subjective meaning: conjectures based on objective meaning-contexts must be formulated about subjective meanings, which then lead on to 'genetic analysis'.

In situational analysis, Popper combines the 'biological approach' to the social world with the assumption that one can discover more through the analysis of the consequences of action, than one can through the analysis of the action themselves. From the formal point of view these two propositions are very similar. Where they differ is in their goal: for Schutz the primary aim is an adequate understanding of subjective meanings, whereas Popper is concerned to establish, as far as is possible, an approximation of the facts in the social world, to the 'objective truth' for means of technological instructions.

Both Popper and Schutz advocate that social scientists should begin with subjects rather than structures, and that the construction of models plays a central role in explaining and understanding actions. Such models are described by Popper as 'nil hypotheses' [*Nullhypothese*] and by Schutz as concepts of 'empty formulae' [*Leerformeln*]. Both mean by this that the models are constructed from ideal-typical statements/assumptions which do not need to correspond absolutely with empirical facts to be successful as heuristic aids. Both derive the necessity for model construction from their basic belief that there is no causal determinism in the social world, and thus that all attempts to formulate causal social laws based on the procedure of the natural sciences are misguided.

As to precision in sociological model formation, both Popper and Schutz point out that models can have variable 'information content' [*Informationsgehalt*] (Popper), or variable 'content-richness' [*Inhaltserfülltheit*] (Schutz). Both agree that the more general the ideal-typical assumptions are, the greater the opportunities for their application. Both agree that, at the same time, the more general the statements, the less the significance of the knowledge gained about an individual fact. The less precise the assumptions are, the greater their range of application, but the less meaningful the knowledge thereby acquired.

In conclusion I should point out that although Husserl and Schutz advocate an adequate relation of scientific action to everyday actions, they do not discuss the effect of theoretical statements on and in the everyday world. Yet the generally accepted results of scientific and other research endeavours are, through various 'popularizers' (teachers at all levels, journalists, etc.)[3] in the course of time taken for granted in the everyday world. If Popper, on the one hand, emphasizes the importance of scientific findings for everyday actions, Schutz and Husserl, conversely, underestimate them.

SUMMARY

The above discussion of what Popper and Schutz have in common is not exhaustive. It indicates that the social research logic of Popper and of Schutz is not exclusive, and that the relevance of either depends on the particular problem in question. As far as social geography is concerned, we can provisionally conclude that solutions to problems should first be sought using the methodology of the objective perspective. If these prove inadequate, the procedures of the subjective perspective should be followed. Of course this hypothesis needs further investigation and empirical research. We shall return to this in the next chapter, after an investigation of sociological theories of action.

Chapter 5

Social theories of action

This chapter continues my attempt at developing an action-based theory of social geography, highlighting subjective agency. This time I shall concentrate on relevant sociological theries of action,[1] aiming to establish more precise categories of description and explanation of human activity than those emerging from my examination of phenomenology and critical rationalism so far. I am concerned here with clarifying the social contexts of human action, and with the conditions under which subjective agency affects the physical world. In other words: I want to try for more finely differentiated categories and action models for the investigation of the social components of the 'three-world-theories' of Popper and Schutz. And, as I said in Chapter 1, I have another aim in considering the extant sociological theories of action. My argument is that, whatever their considerable value and prestige as social theories, they do not give suffcent credit to the complexities of subjective agency. While they think of themselves as 'action' theories, the action they theorize remains largely mechanical. Their 'actors' capacities for creativity or innovation, for changing the world in which they exist, is not something that these 'action' theories explain. This is something of a paradox, for this reason: the emphasis on 'action' is traditionally allied with an emphasis on individualism. It is also allied with a conservative view of social change, despite the individualist rhetoric of many subsequent sociologists and human geographers who claim authority from the action theories I am about to discuss. The fact that the actors in these theories have, in reality, little power to change the order in which they exist resolves this paradox, but it does not explain the fact that subjective agency can effect changes. After discussing the existing action theories, I shall develop this point in more detail in the next chapter. In this chapter, I shall, after dissecting extant action theories, consider how they can none the less be useful, in different ways, in empirical research into the various problematic situations of action.

In the course of the development of social theory, the meaning of the central theoretical concepts of 'action' and 'act' have become considerably differentiated and modified. Action models and action-theory approaches based on such concepts can only, in my opinion, be appraised according to their success in solving practical social problems. As the practical problems confronting geographers (among others) in the ecological area are increasing, this preference for practicality is more than an academic criterion. We have to ask which action model is likely to offer the best solution for which kind of problem, before we can recommend appropriate proposals for solutions. We also have to establish which models are most suitable for research into objective spatial structure, the production and use of the spatial patterns of material artifacts, as well as ecological problems. We are thus attempting to improve the preconditions for research in social geography, by taking more account of theoretical advances in sociology.

Before considering the various sociological action models, we shall first establish a framework for analysis. If we are to compare different action-theory approaches and define their areas of application in detail, they must be analysed in terms of the same categories. In the next section, I shall begin to do so.

ANALYSIS FRAMEWORK FOR THE COMPARISON OF ACTION MODELS

In Chapter 1, I said that the two main defining features of 'action' were reflexivity and intentionality. Throughout the history of the social sciences both features are found in all definitions of 'action' and 'act'. In addition, the processual categories of 'action project', 'definition of the situation', 'realization of the action' and 'consequences of the action' are to be found in every action model. Furthermore, from the structural point of view there are four elements in all action models. These are either implicit or explicit:

– the agent, who has realized the action or intends to do so;
– the goal or the goal orientation of the action, which results from various alternative action projects;

- the frame of reference of the action orientation[2]; and
- the situation in which the action is carried out, and how this situation is defined in relation to the action project.[3]

The different action models ascribe varying degrees of significance to these four factors. The significance attributed to them also varies in empirical research, as do the interpretations of these factors. What is lacking here are grounds for a systematic distinction between these degrees of significance. These I shall attempt to establish.

The various action models can be systemized on grounds established in the last three chapters. Critical rationalism and phenomenology are seen here as extreme poles in the continuum of increasing thematization of social reality. The criterion for the classification of the various action models is therefore the extent to which they are questioning social reality.

My analysis of the sociological background of social geography will focus on three models of action:

1 The purposive-rational model of action, and the related rational choice theory approach which seeks to analyse sequences within this model. Both approaches are applied not only in sociology but also within (neo-classical) economics.
2 The norm-oriented model of action, which developed in the context of structural-functionalist social research.
3 The action model of intersubjective understanding, developed in phenomenology and ethnomethodology (but I shall restrict my investigation to a reconstruction of Schutz's original model).[4]

The *purposive-rational model* represents the least comprehensive approach. Social reality is defined as a given environment and the question becomes: by which means can agents successfully gear into it. Rational choice theory analyses more closely the 'choice' between alternative means within the framework of a given goal. The *norm-oriented model* goes beyond the investigation of goals and the alternative means to achieve them, analysing the various kinds of orientation of actions towards cultural values and social norms. What I regard as the most comprehensive basis for an analysis of the social world, I term the '*model of intersubjective understanding*'. Here not only are the choice of means or norm-reference themes investigated; so too are the constitution of goals and the meaning-contents of human actions. Schutz is concerned to discover the preconditions which make communication possible between the various subjects in a society, and the existence of the social world in general.

The three action models also differ in their implicitly or explicitly expressed concepts of rationality. I am not, however, suggesting that these action models are mutually exclusive. On the contrary, they have produced specific judgement criteria for problematic aspects of human action: for the problem of choice of means to unproblematic goals in the case of the purposive-rational action model and choice theory; for the reconstruction of value and norm-related aspects in choice of goal and means in the case of the norm-oriented model, and for the discovery of subjective meaning-contents in communication difficulties in the case of the intersubjective understanding-oriented model. Which of these aspects becomes relevant in a particular action depends on the nature of the problem to be solved and/or on the theoretical interests of the social scientist involved.

We cannot go into each action model in detail here. As I indicated, the aim of the following systematic analysis is to identify the sociological points of departure in the development of an action theory for social geography in more detail. I shall analyse them in order of their thematic complexity.

THE PURPOSIVE-RATIONAL ACTION MODEL

The concept of purposive-rational action is primarily an ideal-type construction. As a general basic category of sociology it was developed by Vilfredo Pareto – under the name 'logical action'- and by Max Weber. It is also connected with Pareto's theory of 'choice acts' [*Wahlakte*], and is the ostensible point of departure for the development of rational choice[5] and game theory.[6] Since choice theory has been applied in different ways in regional economic human geography, it will be analysed in some detail here in relation to the categories described on pages 101–2 (agent, goal orientation, frame of reference, action situation). We shall first point out the context into which Pareto and Weber placed the model of a purposive-rational action, so that we do justice to the ideas of these two classical representatives of modern social science. At the same time we shall be considering what the two have in common and where they diverge.[7]

Pareto (1980: 35–6) developed a system of classification by means of which actions can be systematically ordered according to their empirically predominant features. The decisive criterion for assigning a certain action to a certain class is the appropriateness of the selection of means for the attainment of a given goal. This selection criterion, however, is in turn subordinate to Pareto's distinction between the 'objective' and 'subjective' aspects of the action. It is from a combination of the two selection criteria, the appropriateness of means and goals regarding (a) the objective and (b) the subjective aspect of the action, that Pareto derives his distinction between 'logical actions' and 'non-logical actions' (Figure 6).

Generas and species	Have the actions logical ends and purposes	
	Objectively?	Subjectively?
Class I: Logical actions The objective and subjective purposes are identical		
	yes	yes
Class II: Non-logical actions The objective end differs from the subjective purpose		
Genus 1	no	no
Genus 2	no	yes
Genus 3	yes	no
Genus 4	yes	yes
	Species of Genera 3 and 4	
	3α 4α The objective end would be accepted by the subject if he or she knew it.	
	3β 4β The objective would be rejected by the subject if he or she knew it.	

Figure 6 Pareto's action typology (from Pareto 1980: 19–20)

In order to show when Pareto sees an action as rational and appropriate to the goal, as a 'logical action', and when he sees it as a 'non-logical action', I must first explain the more important distinction between the 'objective' and the 'subjective' aspects of the action. For Pareto (1975: 58 f.), any sociological phenomenon has two quite separate and often totally different forms: an objective form which establishes relationships between real objects, and a subjective form which establishes relationships between psychological states. Suppose we have a convex mirror. Objects reflected in it will appear distorted: what is actually straight will appear crooked, and small objects will appear large or vice versa. It is the same with objective phenomena as they are reflected in the human consciousness. If we wish therefore to perceive an objective phenomenon, we must not allow ourselves to be satisfied with subjective perception alone: we must deduce the objective from the subjective.

This statement could be understood as a defence of positivism, a demand that sociological analysis follow strict natural-scientific principles. Pareto himself (1910: 234) rejects this notion, however: 'We should not be put off by the names already given to these two categories. They are in fact both subjective, since any human knowledge is subjective. The difference is not one of nature but in the degree of factual knowledge available.' 'Subjective' thus means: 'from the perspective of the person carrying out the action, whose available knowledge is deficient' (Mongardini 1975: 24). 'Objective', on the other hand, means: from the perspective of the scientific observer, who has at his disposal theoretical knowledge which has not yet been empirically disproved.

The two classes of 'logical actions' and 'non-logical actions' should be understood in the above context. Pareto (1980: 19) defines *logical actions* as 'actions that logically conjoin means to ends not only from the standpoint of the subject performing them, but from the standpoint of other persons who have a more extensive knowledge – in other words, to actions that are logical both objectively and subjectively'. The class of *non-logical actions* consists of those actions which choose the (logically) wrong means to achieve a desired goal, from the objective point of view – i.e. from the point of view of an empirically valid theory. However, 'non-logical' is not the same as 'illogical', according to Pareto (1980: 20). For actions shown to be 'non-logical' from the objective point of view can be perfectly logical when seen from the subjective point of view.[8] It is therefore a mistake to describe these actions as illogical. Actions should have specific features if we are to assign them to one class or the other.

For Pareto (1980: 25), 'logical' actions are 'at least in large part results of processes of reasoning'. They derive from a conscious weighing up of the means available for the achieving of a particular goal. Further, the choice of means should prove successful, i.e. the consequences of the action should coincide exactly with the result expected. Although Pareto (1980: 20) starts from the premise that this class of actions is fairly prevalent, he does not think that such actions are instrumental in constituting the social world. The fact that an action is logically correct does not mean for him that it is socially useful. The major and most important proportion of actions which constitute society are in his view non-logical, athough they are nevertheless of immense use in preserving social balance.[9] Accordingly, Pareto's works on sociology are mainly concerned with the theory of non-logical action.

There is an important point about Pareto's definition of the subjective point of view. He speaks of a subjectively correct choice of means, but sees this on the one hand as the result of a (mistaken) psychological state, while on the other he defines 'simply' subjectively correct actions as deviations from logically or scientifically correct action. In this way he judges the closeness of subjective action to logical correctness from the objective perspective. Subjectivity is seen as a deviation from objective, scientifically based 'logical' action. Whether an action is judged to be logical or non-logical depends on the stage science has reached. Basically Pareto takes the perspective of the objective observer, classifying actions according to objective criteria.

Max Weber developed his ideal types of social actions against a background of methodological concepts. The basic criterion for the formation of types is the degree of consciousness of goal orientation on the one hand, and the means chosen for the achievement of the goal on the other. In contrast to Pareto, Weber judges the rationality of a choice of means not from the objective but from the subjective standpoint. This has the advantage of allowing deeper access to the agent's structure of argumentation, with the result that more judicious proposals for improvement can be made on a practical level. However, if attention is only paid to subjective motives and procedures, without any consideration of the action which is 'objectively' the right one, the social scientist remains in an ivory tower.

Weber calls an action showing a high degree of consciousness of orientation a purposive-rational action, i.e. an action 'exclusively oriented towards means which are (*subjectively*) perceived as adequate for goals subjectively seen as clear and unambiguous' (Weber 1951; 428). I should point out that from a methodological point of view, the greater the consciousness of the action's orientation, the greater the evidence for interpretation. 'The highest degree of "evidence" is contained in an interpretation of a rational action appropriate to the goal' (Weber 1951: 430). From these criteria Weber (1968: 24-5) derives four ideal types of social action, of subjective meaning-orientation. Like Pareto, he does not see this classification as exhaustive. He considers it as adequate and useful for his sociological research aims insofar as such abstract types approximate to real action to a greater or lesser degree. Weber's four ideal types are these[10]:

1 *Purposive-rational action* [*zweckrationales Handeln*]. Subjectively rational choice of means for the achievement of goals, i.e. subjectively rational choice of the goal of an action from the point of view of a successful outcome.
2 *Value-rational action* [*wertrationales Handeln*]. A similarly conscious weighing up of the action project. Success, however, is not the decisive factor in the choice of means or of goals. This kind of action 'always involves "comands" or "demands" which, in the actor's opinion, are binding on him' (Weber 1968: 25). The main principle here is thus the observance of accepted values and norms.[11]
3 *Affectual (emotional) action* [*affektuales Handeln*]. Weber (1968: 24) describes this as being on the borderline of a rational orientation. In the context of ideal-type constructions such actions are seen as uninhibited expressions of emotion, or as the 'conscious' discharging of emotional states.
4 *Traditional action* [*traditionales Handeln*]. Weber also describes this ideal type, and the empirically observed actions which more or less correspond to it, as being on the borderline of rational orientation. This applies to all forms of imitative actions which have not, or only perfunctorily, been thought through, which are based on ingrained attitudes and mere habit.[12]

Weber does not maintain that these abstract types occur regularly in their pure form on the empirical level. It is very seldom 'to find concrete cases of action [. . .] which were oriented only in one or another of these ways' (1968: 26). Real actions only approximate more or less closely to these types. More often they consist of a mixture of all four. Ideal-type constructions should always be seen as methodological and heuristic aids, as devices which may or not approximate historical phenomena in varying degrees. Actions can thus be 'identified' in relation to these categories. The meaning of the concept of purposive-rational action should also be interpreted within the context of this methodological objective. The purposive-rational concept plays a central role in the formation of sociological theory and as a methodological aid. But Weber does not maintain that social phenomena occurring empirically are for the most part determined by rational action appropriate to the goal. Like Pareto, Weber (1968: 21) also emphasizes that

> in the great majority of cases actual action goes on in state of inarticulate half-consciousness or actual unconsciousness of its subjective meaning. [. . .] Only occasionally [. . .] often only in the case of a few individuals, is the subjective meaning of an action, whether rational or irrational, brought clearly into consciousness. The ideal type of meaningful action where the meaning is fully conscious and explicit is a marginal case.

Nevertheless, sociological analysis should start out from pure ideal types, in order to be in a position to establish deviations from and approximations to them, and understand and explain actions in a more satisfactory way.

Weber again, like Pareto, also concerns himself with the subjective aspect of action. However, Weber does not consider subjective goals from the objective perspective only. He also examines the agent's 'subjective meaning of actions', and he is looking for an interpretive understanding and explanation of their course and consequences. The task of sociology is to comprehend the process of conferring meaning. It is to establish what meaning the agent attaches to his action. It is not sociology's task to establish what the action might mean objectively or whether, according to objective criteria, it is logical or non-logical. Weber also points out that the subjectively intended goal of actions usually implies a social orientation, and that therefore it need not correspond to Pareto's idea of emotional states.

The subjective interpretation of social meaning-contexts is paramount for Weber, rather than the meaning which a scientific observer 'gives to a process interpreted as an action' (Girndt 1967: 28). In short, *contra* Pareto, Weber seeks to investigate society from the subjective meaning-context. Pareto's 'logical actions' may be characterized as objectively purposive-rational actions, and Weber's 'purposive-rational actions' as subjectively purposive-rational actions. At the same time, both Pareto's 'logical actions' and Weber's 'purposive-rational actions' have the character of models; both writers see them as marginal rather than typical. For both writers, those models are analytical abstractions, equivalent to Schutz's 'bracketing'. Their importance lies in the methodological objectives they establish, and the fact that other ideal types of, and empirical forms of, action are presented, to a greater or lesser extent, as deviations from these models.

Rational choice theory concerns the sequence of purposive-rational or logical actions, a sequence which can be analytically differentiated. It is about 'choice' in various typical situations. Rational choice theory is applied to two different objectives. On the one hand, the aim is to improve the empirical description of actions in various choice situations. This is its descriptive aspect. On the other – and this is its main emphasis – rational choice theory statements should support and advise the agent in his or her efforts to make 'correct' choices. This second aspect may be described as the normative aspect of choice theory, whereby the 'norms' are those of technical rationality, and the correct combination of means and a (given) goal.

It should now be clear that general rational choice theory is primarily concerned with the formal-rational aspect of actions, and not with their empirical-rational nature. In order to achieve empirical rationality, the agent must apply the formal categories of rational choice theory in a way appropriate to her action situation. Theorists of various disciplines or different areas of interest should work out beforehand the appropriate forms of application and adapt them to their particular field. For instance, researchers in regional studies could amplify thematically the formal aspect of general rational choice theory, in such a way that agents have at their disposal clear criteria for choice in their selection of a location for their particular aims.

Rational choice theory has hitherto been applied mainly, albeit in various ways, in the field of economics. But it has also been developed in other disciplines (including political science, law and ethics), where it has become a theoretical basis for empirical research, often in conjunction with the application of technological theory. However the following account will be restricted to the formal aspects of rational choice theory, leaving aside disciplinary differences in its application. At the same time the relation between rational choice theory and Pareto's and Weber's objective and subjective models of purposive-rational action will be borne in mind. All three will now be analysed in terms of the structural framework of analysis outlined above.

Model of the agent

According to Pareto, agents able to design and carry out objectively purposive-rational actions must be familiar with valid empirical knowledge relevant to their goals. They must deduce the means to achieve their goal from the scientific theories available. It is remarkable, I think, that Pareto in the first edition of *Manuel d'économie politique* (1909), defines his theoretical concept in the same way as Popper formulated his own thirty years later: 'We will never forget that a theory should only be accepted provisionally. One we hold true today will have to be abandoned tomorrow if another one which comes closer to reality is discovered. Science is in perpetual development' (1971: 8).

It follows that intervening in the physical word, the agent must be familiar with relevant knowledge from the natural sciences. In other words, an agent able to perform a logical action must be in a position to reflect closely upon the achievement of an intended goal so that 'the subjective fact conforms perfectly to the objective fact' (Pareto 1971: 103). The agent must also perform the action in such a way that there is a logical connection between the relevant knowledge, the objective facts, and the consequences of the action. An agent who is capable of all these things is referred to as 'homo rationalis' in the objective sense.

'Homo rationalis' has the following additional features. He operates on his own, and in the realization of his goals he is influenced only by pragmatic criteria of rationality, having the most comprehensive information at his disposal. He is thus in a position to achieve a correspondence between his own intentions and objective reality, in a way that is logically and empirically correct. 'Homo rationalis' is thus an agent who is totally in control of all the components of the action process.

In *Weber's* view, the capacity for rational thought is paramount for the agent to perform a subjectively purposive-rational action. He or she is thus in a position to weigh up consciously various possible goals on the one hand and a particular goal, the means to it, and consequences of it on the other. In doing so he or she is influenced neither by ethical, moral or other values, nor by emotional elements or traditional attitudes. The only thing of interest is the successful realization of the goal. Unlike Pareto, Weber does not endow his 'homo rationalis' with true knowledge, but 'only' with knowledge which is subjectively held to be true. The most salient characteristics of such an agent are thus a high degree of reflexivity at the point when means are chosen and the action performed, and formal rationality. The agent sets goals based on the subjective knowledge available, choosing the (subjectively) best means to achieve them. Thus the agent projected by Weber should be seen as a 'homo rationalis' in the subjective sense, acting correctly from a subjective point of view, and using available knowledge in the best way.

The model of agent in the *rational choice theory* is endowed with 'attitudes', which are seen as principles of action. 'If these principles are well known and can be formulated mathematically, they simply represent functions which characterize the conduct of an agent or a number of agents' (Gäfgen 1974: 19).

The function which characterizes the agent is the basic criterion of choice theory: it is subjective rationality. With which choice of alternatives can the agent realize personal goals, on the basis of the personal knowledge available? Höffe (1980: 24–5) summarizes the 'subjective features' of choice theory's ideal-typical model of the agent, as follows. The agent must have (i) a conscious goal; (ii) personal 'knowledge of the spectrum of alternatives' (choice of means), 'whether such knowledge is complete or incomplete, correct or incorrect'; (iii) personal knowledge of the results or consequences of the various alternatives available to him or her, 'again, it is immaterial whether such knowledge is mistaken or not'. Finally, agents in a choice situation must be (a) aware of their many values and goals; their values and goals must be (b) so ordered that no contradictions occur, and (c) they must already have found a clear preference.

Goal orientation

Pareto, Weber and choice theory agree to a large extent on the aspect of goal orientation. As I have indicated, *Pareto* (1971: 103) starts out from the idea that all logical actions are directed by the desire to attain things which satisfy the agent's needs. General goal orientation can thus be defined as 'the objectively successful satisfying of needs'.

In *Weber's* (1968: 24–5) ideal-type construction, the aim of purposive-rational action is 'success', the satisfying of a 'subjective interest', meaning the achievement of self-interest. Goal orientation is further determined by relevant feelings of need, which are consciously placed by the agent on a 'scale in such a way that they are satisfied as far as possible in order of urgency, as formulated in the principle of "marginal utility"' (Weber 1968: 26). Subjectively perceived degrees of urgency regarding the satisfaction of needs are thus paramount for goal orientation.

'Generally speaking, *rational choice theory* does not deal with goals, but with the agent's "rank ordering" of goals' (Gäfgen 1974: 28). The idea is that the agent knows his goals, or that the goals are given. The question then becomes: how can the agent achieve a goal with maximum profit/satisfaction? In addition it is assumed 'that the deciding agent is aware of his various values and goals, and that he puts them in rank order, eliminates possible contradictions, and finds a clear goal-(preference)-function' (Höffe 1980: 25).

Frame of reference

Models of purposive-rational action are based on a homogeneous goal structure. Frames of reference are therefore relevant not for goal orientation but for the choice of means for the achievement of given goals. We now see that these models are primarily instrumental in character. Pareto developed a normative frame of reference for this, which the agent must have if an objectively correct action is to be rendered possible. Weber, on the other hand, emphasizes the descriptive aspect by reconstructing the way in which rational actions appropriate to the goal are oriented in an ideal-typical way. I should point out that choice theory is in fact a synthesis of these two models, although 'choice theory' is often ignorant of its antecedents. First an attempt is made to describe subjective knowledge. Then a formal calculation is made as a grid of orientation. This grid may be applied both on the descriptive–explicative and the technical–normative levels.

There is virtually nothing to say about *Pareto's* frame of reference for objectively purposive-rational actions that has not already been said. Enough for here, that this frame corresponds to the relevant accepted empirical theories of the natural and social sciences.[13] If the actions are nevertheless unsuccessful, the scientific law on which they were based should be considered refuted. The correct means should then be sought, through rational trial and error.

For *Weber* (1968: 71ff.), the frame of reference for orientation of the choice of means in subjectively purposive-rational actions is always the subjective knowledge available to the agent. Conscious reasoning, subjectively perceived by the agent as correct, relates to the choice of means for the achieving of given goals. In this sense, the frame of reference as perceived subjectively by the agent may show various degrees of approximation to the objectively correct frame of reference. The degree of approximation thus also represents the degree of likelihood that the action will succeed. A rational orientation is evident, when it consciously and purposefully applies the available knowledge in an attempt to achieve a given goal, whatever the nature of such knowledge.

Thus in *rational choice theory*, the frame of reference refers to the agent's level of knowledge and value system, and to the choice principle for various situations which, on the theoretical level, are ideal-typical projections for practical action. It is in this way that the subjective and objective components are combined. Starting from the agent's subjective knowledge, the 'scientist' should try to help the agent decide which choice should be made by outlining a 'choice grid'. A choice grid consists of various decision possibilities. These should be adapted to the agent's level of knowledge and value system.

Depending on his or her level of knowledge (ranging from absolute to tentative), the agent is aware to varying degrees of the different available alternatives. The same is true for the expected consequences of the projected actions. The agent must now choose between the existing alternatives, and for this he or she needs rules for evaluation (a value system) and for decision (decision principle). The value system consists of coordinated rules, according to which the elements of the situation and the alternatives are evaluated. The decision principles can be described as a rule for the choice of means. The principles in everyday life are usually of a mechanical kind, i.e. those of routine or rules of thumb. In the case of purposive-rational actions, such choices are replaced by rational decisions which relate to more precise principles.

The level of information, the value system and decision principle are in various ways dependent on one another. The decision principle can only be as precise as the ordering of alternatives affected by the value system. If the value system comprises only the scale 'good–bad', the principle cannot demand more than: 'choose the best alternative'.[14]

Rational choice theory examines the relationship between the respective level of information and the corresponding decision principle in particular detail. In a situation where there is 'a decision made on the basis of absolute conviction', where the agent is aware of all the alternatives and can evaluate clearly and predict the consequences exactly, the (normative) principle is: 'maximize your advantage!' In such a situation, the consequences of any selected alternative can be measured (in units of money, time, etc.), so that with the help of a matrix of usefulness they can be evaluated on a ratio scale. The agent should decide on the alternative whereby the consequences produce a greater advantage (by x units) than any other.

In a situation of 'decision where there is risk', where the agent can only depend upon conjectures, knowledge of the consequences of the various alternatives in choice of means is accordingly only based on probability. It might be possible to maximize benefit, but the decision principle cannot be clearcut. The evaluation of various alternatives must resort to a matrix of usefulness, which consists of a (nominal) scale of the kind 'bad–mediocre–good'.

In a situation where there is a 'decision in the face of uncertainty', the agent is not even aware of the probable consequences of the various alternatives. In that case even the evaluation cannot be indicated: perhaps one should be pessimistic, perhaps optimistic. The evaluation is neutral.

Depending on the degree of uncertainty, the theorist can give the agent a decision principle which is either neutral, cautious or encouraging. Advising caution means minimizing the disadvantages in the case of an unfavourable situation, and 'the choice of the alternative whose worst consequences [are] still better than the worst consequences of the other alternatives' (Gäfgen 1974: 382). Recommending risk often means maximizing the advantages of a favourable situation, and 'the choice of the alternative which is better than all other alternatives in regard to the result for which that alternative is best' (Gäfgen 1974: 381).

The action situation

Pareto defines the action situation of 'homo rationalis' in relation to the goal of the action, where the complete information at the agent's disposal is used in the relevant context. The agent's interpretation of the action situation coincides exactly with the 'if' component that Popper regarded as the primary condition of the scientific law. If the situation is defined correctly, i.e. if the definition coincides with the true facts and the 'if' components, then the logical action is performed with the appropriate choice of means in such a way that the consequences predicted by the 'then' components of the law actually take place. The situation is thus defined in relation to the goal, on the basis of hypothetically true knowledge. As a result of the definition of the situation, the agent finally chooses the appropriate means and performs the relevant action.

In the case of animals, Pareto (1980: 25) postulates a direct relation between 'hypothetical circumstances A' and activities B. In the case of human beings, on the other hand, this psychological state is manifested not only in 'activities B' but also in a third category of expressions C, which in his view frequently appear in the form of theories.[15]

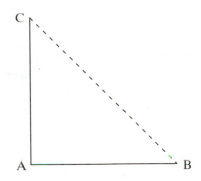

Figure 7 Situation definition in the case of non-logical actions

Pareto applies the schema in Figure 7 to non-logical actions only. But in the light of his distinction between logical and non-logical actions, we can also apply it to logical actions. A logical action B is performed on the basis of the fact that A and C are congruent, that C represents an empirically true theory, and that the latter can be applied appropriately to the action situation in question. From C, therefore, the means can be deduced which are a component of the action situation and which must be selected for the achievement of the goal, if through the action the intended and objectively correct consequences are to be brought about.

Concerning the action situation, *Weber* (1968: 7) maintains that all supposedly 'processes and phenomena devoid of meaning', from the point of view of their relation to human action, should be seen

> in the role either of 'means' or of 'end'; in a relation of which the actor or actors can be said to have been aware and to which their action has been oriented.

However,

> all processes or conditions – whether they are animate or inanimate, human or non-human – are in the present sense devoid of meaning insofar as they cannot be related to an intended purpose. That is to say they are devoid of meaning if they cannot be related to action in the role of means or ends but constitute only the [. . .] favoring or hindering circumstances.

These aspects of the action situation should be seen as 'data' for the agent, as conditions of the situation. They may, of course, have some influence on the form of the action, but they cannot become means. This is either because they lie outside the agent's reach, or because they have no significance in relation to the goal of the action.

Agents use only those elements of the situation which appear from a subjective standpoint, and to which they have access, as means to achieve their goal. In this definition of the situation, however, agents do not allow themselves to be affected by emotions, nor by ill-considered adherence to traditions, but are solely directed by the 'subjective ends' of their actions. They relate their choice rationally (a) to their expectations of the behaviour of objects in the external physical world, and (b) to the expectations of other people in their social environments. In the action projection, such expectations are defined as the 'condition' of the action, or they are integrated into the process of action as 'means' 'for the attainment of actor's own rationally pursued and calculated ends' (Weber 1968: 24). In other words, the agent consciously defines the elements of the situation with the help of the subjective knowledge available, and with regard to a subjectively perceived goal. In Weber's view, the definition of the situation can be described as purposive-rational, as soon as the agent deciding on the means which represent the elements of the situation is directed by nothing other than a subjective intention to realize a goal. That is to say, the agent is influenced neither by ethical or moral values and precepts, nor by feelings or the argument that 'it has always been done this way'.

Similarly, in *rational choice theory* the 'situation' is subjectively defined as that portion of the world that appears relevant to the agent in relation to a given action. The elements of the situation are divided into (a) the subject performing the action and (b) his environment, which is composed of physical and social entities. The situation is always limited by the available knowledge, the frame of reference. This knowledge relates, on the one hand, to the initial situation (I) and, on the other, to the expected consequences of the action (situation II). In order to be in a position to achieve the goal, the agent must be able adequately to assess both situations. Three types of decision situation can be distinguished: situation of certainty, situation of risk and situation of uncertainty.

FUNCTION AND ACTION IN SOCIAL THEORY

The famous so-called action theories of structural-functionalism (cf. Parsons and Dahrendorf) are based on a norm-oriented concept of action.[16] These approaches place the social order and the significance of norms in the forefront of their analyses of actions or society. They originate in funcionalist sociology, represented especially by Durkheim. These 'action' theories of structural-functionalism are, in fact, concerned with only two types of actions. On the one hand, actions relating to 'the objective world of material phenomena', and, on the other, actions relating to the social world. The agent belongs to the social world as a subject interpreting his or her role, in the same way as 'do additional actors, who can take up normatively regulated interactions among themselves' (Habermas 1984: 88). The unity of a social world is delimited by the scope of the validity of its value standards and norms. Within their area of validity, values and norms stipulate which forms of interpersonal relationship are legitimate, and which means should be selected for which goals.

The concept of 'function', the central idea behind this approach, is the precondition for an adequate understanding of norm-oriented social theory. The concept of function from

Durkheim (1984: 11ff.) via Malinowski (1990: 3ff. and 36ff.) and Radcliffe-Brown (1952) to Parsons (1964: 30ff.; 1975), Merton (1957: 30ff.) and Gouldner (1973: 369ff.) is: 'Function' is based initially on an analogy with a biological organism (cf. Spencer 1966/7: 197ff.). Every part of the organism has a function for the organism as a whole. (No lung, no organism.) By an analogy between organisms and the social totality, every part of the totality can be considered either functional or dysfunctional. Hence the notorious history of 'dysfunctional' elements being regarded as deviant, insofar as they disturb the 'equilibrium' of the social whole as it stands.[17]

A differentiation is also made according to the intention with which the action is performed. A distinction is made between those cases 'where the subjectively intended goal coincides with the objective consequences, and those where the two diverge' (Merton 1957: 34).[18] Actions in the first category are termed manifest functions, and those in the second, latent functions.

The question of course is this: Should we be considering from the functional/dysfunctional perspective the *effects* of a certain action, or the project underlying the action? With the exception of Parsons (although Parsons himself does not make this clear), all the writers mentioned above concentrate on the effects of actions. Effectively, Parsons also takes the projects of actions into account. He does so because he incorporates the normative aspect of action (on which functionality/dysfunctionality depends) into his theory of 'pattern variables'. For Parsons not only 'action', but 'system' is a key concept. While in his early work, he regarded a set of 'unit acts' as a system, and assessed consequences of acts according to functionality or dysfunctionality, Parsons' later work is more complex. From 1951, he saw *one* act as a system, whose orientation and projection might be functional or dysfunctional according to the subsystems 'personality' (order of needs), 'society' (social norms) and 'culture' (cultural values). Münch maintains that Parsons retained this view, modifying it constantly with new differentiations, until his last publication in 1978. 'Each individual action is now seen as a product of the interpenetration of these subsystems' (Münch 1987: 35). It is for these reasons that I think Parsons gives more tacit credit to underlying subjective motives than other functionalists, but this does not eliminate other limitations in his understanding of the agent, or of action.

Model of the agent

For Parsons, the agent is a member of a social group. One is primarily a member of a social group. Society is no longer just an environment. Parsons' 'homo sociologicus'[19] is equipped with a motivational complex and the capacity to internalize norms and values, and interprets personal needs in relation to them.[20] The agent's action projects and their results are then either functional or dysfunctional, regarding the interaction between personal needs and social norms. If one needs a lot of money quickly, one is more likely to perform a socially dysfunctional act. But the agent's *rationality* is judged not only on whether he selects the appropriate means for his projected goals, but above all on whether or not his actions are normatively correct and his chosen norms socially valid.

Goal orientation

So the goal orientation of action is controlled by adherence to values and norms, the normative context. Accordingly, Parsons insists that the means for achieving a goal cannot only be selected in terms of their technical effectiveness. The social, normative components also have to be analysed. When we ask whether or not the choice of means is functional or dysfunctional, we ask if it is justified or desirable. The goal orientation of action should be seen 'as the result of an interpenetration[21] of means–end rationality and a normative limitation on the free play of such rationality' (Münch 1987: 17).

There are two other key points about Parsons. Despite his functionalism, Parsons believes that the nature of the orientation of the action creates a social order which can only survive as long as agents *voluntarily* (and not because they are coerced by authority) accept a normative frame of reference. He believes he has developed a theory of a voluntarist social order, together with a voluntarist action theory. He believes this in spite of the funda-mental role social norms play in his theory: 'On no account, however, does the voluntarist system deny the significant role played by conditional and other non-normative elements. They are regarded as being interdependent with the normative elements' (Parsons 1937: 82).[22]

The second point about Parsons is this: actions, and action situations, are necessarily perceived within normative limits. In addition, this normative framework of orientation must have a validity which is independent of the particular situation in which the action takes place. Thus a certain consistency in the goal orientation of action becomes possible, and similarly the agent can have justified expectations about other agents' actions. At the same time, and paradoxically, Parsons argues that the very thing that makes autonomous goal orientation possible, in spite of the agents' 'solidarity' obligations, is also the thing that shapes these obligations. Both depend on the internalization of normative action patterns through socialization processes.[23]

Frame of reference

Thus far, we have seen that the frame of reference for the orientation of an action in Parsons' model depends on the expectations of the other members of the social world, who in turn orient their actions according to common cultural values and social norms. Agents try to fulfil these expectations in order not to jeopardize their 'social membership', or put the social order at risk. In *The Structure of Social Action* (1937: 44), Parsons explains the significance of this frame of reference for his theory:

> The means used within the control possibilities of an action should be seen neither as purely arbitrary in their selection, nor as totally dependent on the *conditions* of the action. We must rather consider them as being *directed* by an independent, quite specific normative selection factor, which we must be aware of if we are to understand a concrete action process'.[24]

As we have seen, Parsons takes the view that agents are not only oriented towards expediency, but also towards social norms, although he does not specify exactly how this orientation operates in individual cases. In *The Social System* (1952), the frame of reference is divided into three subsystems:

- cultural system (symbols; value judgements)
- social system (norms; adaptions)[25]
- personality system (order of needs; motivational orientation).

Every act which can be empirically established is now seen as the result of an orientation in which these three subsystems mutually interact. This means that actions are not oriented, as hitherto, only towards social norms, but also towards cultural values and an order of needs. All three interact. The reproduction of social order and culture by means of actions depends thus on an adequate 'adaptation' of personal needs to the cultural and social systems. Mutual fusion of the three subsystems is functional when purposive-rational action interacts with normative obligations, and when the personality has internalized the normative culture.[26] In this way a basically consistent pattern is established, which Parsons terms 'pattern consistency'. Pattern consistency and its retention is the precondition for the maintenance of a 'voluntarist' social order, although the central element in this 'voluntarist' order is still the solidarity of all the agents belonging to it.[27]

It should be borne in mind that Parsons sees each of the subsystems as *aspects* of the action orientation. 'Which subsystems and which system levels determine action to what extent, and how these factors relate to one another, is not a question which can be decided *a priori*, but an empirical one dependent on varying preconditions' (Münch 1987: 127). In order to specify these conditions, Parsons defines the frame of reference of action orientation more precisely. His key question is how agents can orient their actions in *concrete* situations in such a way that the expectations of the other agents are fulfilled, and a Parsonian voluntarist social order made possible.

How do agents apply normative orientation in their actions, i.e. how do they solve the dilemma of choice in a concrete situation? The answer is: They orient themselves towards the various alternatives allowed by the objects of the concrete situation and the 'pattern variables' (Parsons 1952).

The objects in turn are further differentiated into 'social objects' (the other agents) and 'physical objects'. The perception and experience of both kinds of object can occur in universalistic *or* particularistic orientations, and in performative *or* qualitative orientations. Through differentiation it is thus seen that action always takes place in a certain situation which is composed of objects. 'Attitudes' classify which affective (or 'cathectic'[28]) and evaluative relations an agent has towards the objects in the situation in question. The agents' attitudes in their relation to their objects may take the form of specific *or* diffuse orientations, affective *or* affectively neutral orientations. Parsons uses the concept of 'attitude variables' to express the idea that actions always relate to an agent's particular orientations.

The second important step in the differentiation of the orientation framework of actions is relating the pattern variables (Parsons *et al.* 1953) to functions of action orientation. The four analytically derived functions are defined in this way. (This definition is often referred to as the AGIL schema[29].)

A '*Adaptation*': adaptation to external circumstances by means of instrumental activity (personality system).
G '*Goal-attainment and goal-selection*': adequate action orientation and realization with the goal of maximum satisfaction of needs (personality system).
I *Integration*: attempt at the highest possible integration of action into the specific expectations of the other agents (social system).
L '*Latent pattern maintenance*': adequate reinterpretation/ reproduction of value-symbolic orientations, i.e. maintenance of the action patterns (cultural system).

In devising a differentiated analysis grid for the empirical investigation of action orientations in concrete situations, this four-function schema is now combined with the pattern variables, resulting in the relationships shown in Figure 8.

A Adaptation function	Function of goal-selection/attainment **G**

attitude
 	
  specific
  affectively neutral
 	
object categorization
 	
  universalistic
  performance oriented
	attitude
	
	  specific
	  affective
	
	object categorization
	
	  performance oriented
	  particularistic

attitude
 	
  diffuse
  affective
 	
object categorization
 	
  particularistic
  qualitative
	attitude
	
	  affectively neutral
	  diffuse
	
	object categorization
	
	  qualitative
	  universalistic

I Integration	Latent pattern maintenance **L**

Figure 8 Orientation alternatives

Finally Parsons (1967) narrows his definitions even further by making a distinction between 'the axis between external and internal orientation, and the axis between instrumental and consummatory orientations' (Münch 1987: 46). Here, Parsons reverts to the distinction between conditions and means, and between normative orientation and the attainment of the goal as an end in itself.

> Externally oriented action is determined by outside conditions, as opposed to the internal normative orientations. The consummatory orientation means that a goal would be sought for its own sake, while the instrumental orientation implies that the action is engaged in as the means toward a further set of ends.
>
> (Münch 1987: 46)

The action situation

As Jensen (1980: 58) points out, the agent perceives the specific meaning of the situation, according to Parsons (1952), in the following terms:

(i) What objects are at the base of the situation (from the cognitive standpoint)?
(ii) What meaning does the situation have, with its corresponding objects, in relation to the desires and needs of the agent? Are there any correspondences and/or contradictions?
(iii) Which evaluation of the situation should the agent adopt? Are there any (social) restrictions, or is he or she free to realize personal desires?

For Parsons, (i) and (ii) only establish the facts of a given situation in purposive-rational action (although in purposive-rational action, expediency calculations take the place of desires and needs). But the last point, (iii), points to the norm-oriented goals of actions and the frame of reference. This orientation involves balancing the first two aspects of the situation ((i) and (ii)) in relation to the cultural and social contexts.

Finally, and crucially, Parsons (1952: 53ff.) assumes that a situation is made up of *external* circumstances. 'The situation comprises physical, social, cultural and also organic objects' (Münch 1987: 33). And the objects that matter are only those objects which form part of the agent's situation. Whether these objects are means or conditions for the action depends on whether they can be controlled. This emphasis on control itself signals that while Parsons had some working concept of motivation in function, and a more developed one than his predecessors, it was still a concept that defined action in relation to functional, rational ends. One could argue that 'intersubjective communication' mattered for Parsons, in terms of his normative understanding of a voluntarist social order, but there is no account of how subjective agency is formed in relation to the agency and actions of others. Parsons' actors are uniform, in that he does not account for their particularity, nor is the fact that subjects communicate from their own particular standpoints investigated. His agents in the social order do not interact with one another in a way that permits their subjective agency to affect that order. For that, we need to return to Schutz.

ON INTERSUBJECTIVE UNDERSTANDING

Yet the moment we return to Schutz in search of an action-oriented, intersubjective communication theory, we encounter an obstacle. Schutz did not complete his theory of action.[30] It is something that needs to be developed. To this end, we need, first, to examine the beginnings of the theory of social action he sets out. Second, we need to relate Schutz's gestures towards an action theory to the model of intersubjective understanding outlined above. The following questions have to be addressed here: What leads Schutz to maintain, within the framework of his sociological writings, that society can be analysed from the subjective perspective? From the point of view of action theory, what are the bases on which members of a society can communicate in an intersubjectively valid way? On the theoretical level, what are the conditions under which social phenomena are taken into account from the agent's point of view? How can the individual agent communicate with others about the construction of social reality?[31]

Model of the agent

For Schutz (1962: 76) the agent does not relate directly to the objective physical and social worlds. He points out straight away that during the course of every action the agent is in a particular, biographically determined situation and dependent upon the personal stock of knowledge at hand, the meaning of which we have discussed in Chapter 3. Therefore, any agent engaged in any action is an historically specific, unique subject. All agents are distinguished in their own life-worlds by their specific positions in the social world, the particular constitution of their stocks of knowledge, their current stage in their life-cycles, and their spatial position in the physical world. Whether an agent is married, single, divorced or widowed, whether he or she is a father or mother, or in employment or not, leads to various horizons of experience which take on importance in the constitution of the agent's life-world.

The physical, spatial position of the agent's body takes on particular relevance. But the idea of 'spatial position' cannot be reduced to physical measurement alone. In their analysis of the 'spatial divisions of the everyday world', Schutz and Luckmann (1974: 36ff.) distinguish between the 'world within actual reach' and the 'world within potential reach'. For Schutz, the world within actual reach consists of those sections of the world which can be known through direct experience. The nature of the world within actual reach at the time of the action, and the nature of the world within the potential actual reach, are no less important for the constitution of the agent's stock of knowledge than the stage in the life-cycle.

As I discussed in Chapter 3, *rationality* features in Schutz's model of what I have termed 'intersubjective understanding'. But Schutz's rationality does not only relate to the degree of goal-attainment as it does in Weber's theory. Schutz's rationality also bears on the conditions under which the intersubjective world can have validity as such for the community of agents. Technical or instrumental knowledge ('know-how'), the awareness of the validity of norms and values, and knowledge of the significant/relevant circumstances all have to be taken into account if agents are to arrive at an intersubjectively shared definition of the situation. In the intersubjective model, actions are accordingly rational if they 'have the character of meaningful actions intelligible in their context' (Habermas 1984: 13), and if, through them, the agent is relating to something in the world, shared with other agents.

Goal-orientation

For Schutz, the analysis of the goal orientation of action is the same as the investigation of the process 'by which an actor in daily life determines his future conduct (activity) after having considered several possible ways of action' (1962: 67). To deal with the goal-orientation in the subjective perspective, Schutz found it necessary to differentiate clearly between terms which had hitherto been considered synonymous.

But let us return to Schutz's preliminaries to a theory of action in the model of intersubjective understanding. He defines 'action' as a process of human activity based on a previously designed project. An 'act', on the other hand, describes the outcome of the action, the ongoing process. Schutz does not consider every projected activity as an intended activity, which leads to a further important distinction. 'In order to transform the forethought into an *aim* and a project into a *purpose*, the intention to carry out the project, to bring about the projected state of affairs, must supervene' (1962: 67). The decision 'let's begin!' should be interpreted as the result of deliberation 'whether or not to carry out a projected action as the choice between [. . .] projects, [. . .] anticipated states of affairs [. . .] as a dramatic rehearsal in imagination of various competing possible lines of action' (1962: 68).

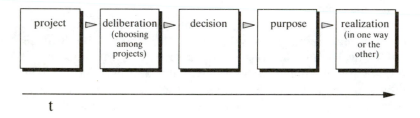

t

Figure 9 Process of goal orientation

In this context, Schutz asks whether projects which always relate to the 'phantasying' anticipation of future activity are concerned with 'action' or 'acts'. For him there is no doubt that any projection relates to the performed act. It is only from the standpoint of the imagined end product that the agent can decide on the separate steps which are supposed to lead to it (see Figure 9). The agent has to place himself or herself in fantasy 'at a future time when this action will already have been accomplished, when the resulting act will already have been materialized. [. . .] What is thus anticipated in the project is [. . .] not the future action, but the future act, and it is anticipated in the Future Perfect Tense, *modo futuri exacti*' (Schutz 1962: 69). This curious temporal perspective has important consequences.

The agent's fantasized idea of the desired final circumstances cannot determine what will actually happen. The fantasy can only be based on knowledge which the agent has of the results of previous acts which are similar in type to the present act in question. At the same time the agent must have knowledge of 'typically *relevant* features of the situation in which this projected action [sic!] will occur' (1962: 69). Thus the agent must be able to recognize those elements of the action situation relevant to attaining the aim, and be able to make a choice. Schutz emphasizes that while projecting an act the agent has only a stock of knowledge when he or she is fantasizing the circumstances he or she wishes to bring about. Therein lies the basis for 'the intrinsic uncertainty of all forms of projecting' (1962: 69). The basic categories for the factors determining motives and goal orientation can be derived from this temporal perspective on action projects. In other words, the temporal perspective provides the clues for answering the questions of whether goal orientation (the 'in-order-to' motive) is determined by the stock of knowledge (the 'because' motive) or whether, in spite of the stock of knowledge's central importance, the agent can direct his or her own goal orientation regardless.

A key ingredient in all action theories concerns the occasion of the action and the projection of the act. Schutz attempts to answer it from the subjective perspective. For him, every goal orientation is connected with the agent's stock of knowledge, since it always bears on previous acts which are similar in type. More specifically, goal orientation is connected with the way in which the agent relates to that knowledge at the time of the projection of the act. And 'this knowledge is an exclusively subjective element, and for this very reason the actor, as long as he lives in his projecting and acting, feels himself exclusively motivated by the projected act in the way of "in-order-to"' (Schutz 1962: 72). Thus every human action is shaped by the agent's projects. The project constitutes the primary meaning of the action. 'In other words, the act thus projected in the future perfect tense and in terms of which the action receives its orientation is the "in-order-to motive" for the agent' (Schutz 1972: 88). It follows that the component representing a genuine 'because' motive is the field of possibilities within whose framework action projects are possible. Because the agent, up until the time of the action project, has had certain experiences and not others (i.e. he or she has only limited knowledge of acts which are similar in type), the choice of action projects is limited accordingly. Thus the agent's previous history limits the field of possibilities for action projects. However, in the action situation at the time of the project, the agent refers subjectively to the stock of knowledge which includes experiences of action situations of a similar type. Since he or she refers subjectively to this knowledge, the agent is not aware of the 'limited nature' of the possibilities. From the subjective standpoint, it is thus always an 'in-order-to' that directs the action project and, in the end, the action itself, just as it also constitutes the action's meaning.

Which standards does the agent take as a guide when choosing among several possible projects? For Schutz (1962: 84), action theory approaches have hitherto implied that during the deliberation stage agents have to choose solely among pre-set, well-circumscribed problematic alternatives, just as the agent can decide among pre-constituted means for the achievement of a goal. These schools of thought, best exemplified in the purposive-rational and the structural-functionalist theories discussed above, maintain that the choice of an action project proceeds in the same way as the choice between means A, B, C or D. Schutz sees this as an abbreviated version of the problem of 'goal-orientation of the act'. In his view all possible projected alternatives depend on the (biographically) determined situation of the agent, on the particular form which the agent's stock of knowledge takes at the point in time when the project is conceived. The projects are however not pre-given, but are only constituted in the subjective light of the stock of knowledge ('in-order-to' motive). In contrast to the choice between the pre-given means of an action, 'projecting is of

[the agent's] own making; [. . .] there are at the time of [. . .] projecting no problematic alternatives between which to choose. Anything that will later on stand to choice in the way of a problematic alternative has to be produced [by the agent], and in the course of producing it, [he or she] may modify it at [his or her] will within the limits of practicability.

(Schutz 1962: 84)

In the subjective perspective, 'the ego living in its acts knows merely open possibilities; genuine alternatives become visible only in interpretative retrospection, that is, when the acts have been already accomplished' (Schutz 1962: 87).

This description of goal orientation is a hypothetical foundation for my insistence on the idea that there are no meanings independent of the subjective agent. It is the agent who reconstitutes meaning with every action project. These are thus not totally determined by the nature of the stock of experience, but only limited by it. This standpoint impicitly refers to, and introduces, the 'frame of reference' in action orientation.

Frame of reference

It is necessary now to clarify which 'scheme of reference' (Schutz 1962: 7) is used by the agent in an action orientation, in a choice of project. This 'mind's selectivity' (Schutz 1970: 5), or consciousness, can be given a theoretical similar analysis. As title for the 'basic phenomenon we suggest "relevance"' (Schutz 1970: 13).

From the analysis thus far of the model of intersubjective understanding, it will be clear that the specific state of the agent's stock of knowledge in a given biographical situation forms the reference scheme, i.e. the frame of reference for action orientation.

Again following William James, Schutz (1966: 120–1) makes a fundamental distinction between two forms of knowledge which constitute the agent's stock of knowledge, and which should be treated separately in the analysis of action orientation:

1 '*Knowledge about*'. This refers 'to that comparatively very small sector of which everyone of us has thorough, clear, distinct, and consistent knowledge, not only as to the what and how, but also as to the understanding of the why, regarding a sector of which he is a 'competent expert' (Schutz 1966: 120). 'Knowledge about' thus consists of specialist knowledge, for instance the expertise required of an electrician, doctor, etc.[32] The agent in this position can analyse objects unambiguously and adopt instrumentally rational orientations.

2 '*Knowledge of acquaintance*'. This comprises the main part of the stock of knowledge, and allows agents to orientate themselves appropriately to everyday life without having to be an expert in these areas.[33] 'It merely concerns the what and leaves the how unquestioned' (Schutz 1966: 120). The types included in this knowledge are to some extent incomplete. Many aspects of the life-world only become problematic as soon as they suddenly

acquire importance *via* an 'in-order-to' motive. Before, they were bracketed.

Thus, in speaking of the two areas of knowledge, 'it is only the "knowledge about" that stands under the postulate of clarity, determinateness and consistency' (Schutz 1966: 121). 'Knowledge of acquaintance', in spite of all its contradictions and inconsistencies, is '"taken for granted", as long, at least, as such knowledge suffices to find through its aid one's way in the life-world' (Schutz 1966: 121), as long as the types and typifications allow for adequate action orientation. These types can be seen as the basic form of everyday knowledge in action. They may be defined more precisely:

1 The type which is a sedimentation of experience determined by situations aids action orientation in the situations for whose interpretation it was evolved.
2 The fact that it aids orientation indicates that it must prove itself in situations which are unclear, i.e. that it must mediate between certain and uncertain elements.
3 A temporal aspect can also be derived from this: 'Although this type is bound up with certain contents linked with the past, it carries with it expectations relating to an unknown future, which anticipate courses of action but are surrounded by a bevy of alternative possibilities' (Srubar 1979: 45).

Yet for intersubjective understanding concerning action orientations to be possible, further conditions have to be fulfilled: 'type adequacy' and 'reciprocity of perspective'.

'*Type adequacy*' means that the precondition for any understanding (between agents on the one hand, and between agents and social scientists on the other) is that the types of stocks of knowledge must coincide with the constructions of the other agents in such a way that they are as comprehensible to the agent as to the agent's contemporaries. 'Adequacy' is achieved when people have experienced the same series of types and are confronted with comparable areas of experience. This shared frame of reference is the necessary condition for intersubjective agreement regarding action orientations.

The possibility of mutual deviation from types, on the basis of the biographically specific nature of the stock of knowledge, is, in Schutz's view, excluded by agents in everyday life because of the 'general thesis of *reciprocity of perspective*'. This implies that 'I and my fellow-man would have typically the same experiences of the common world, if we exchanged places' (Schutz 1962: 316).[34] This thesis is thus based on the assumption that all agents live in a commonly shared (social) world in which they share communal experience about what happens, and that they act as if the experience perspective of others would also be accessible to them if they tried to put themselves in the other's position. Only then can the meaning-constitution of the act (the 'in-order-to' motive) take place in a mutually comprehensible way, and make mutual agreement about action reference points possible.

Still it remains unclear in which circumstances, and in what way, the agent refers to this shared scheme of orientation. In other words: which aspects, in connection with pragmatic motives of action, are relevant for adequate orientation?[35] When does the obvious become a problem and the action orientation problematic? I shall attempt to answer these questions with reference to the following example taken from Schutz:[36]

> It is winter. A man approaches a house situated in the middle of a mound of stones, and enters a dimly lit room. He sees something coiled under the bed. Is it a coil of rope, or is it a snake? He has to make sure what it is. Because he wants to go to bed soon, without being bitten by a snake, he picks up a stick and prods the thing under the bed to see what happens.[37]
>
> Yet it is this coil under the bed that is important for the man. There are other objects in the room which he enters, but they are not of the same importance to him. They all remain within his unstructured field of vision, however.

Against the background of the stock of knowledge, all immediate experiences 'are matched with or superimposed upon the types of the already experienced material' (Schutz 1970: 22). Thus the man can 'recognize' immediate experience. He can tell whether or not what he is experiencing agrees with the types with which he is familiar. Such overlaps remain in the unstructured field, in the background. Problematic experiences stand out from this background and become important, or thematically relevant. In other words, consciousness executes a thematic selection procedure. In this form of reference to the stock of knowledge as a scheme of action orientation, certain elements become the thematic core, within the framework of a certain horizon of perspective.[38]

The problematic object is now thematically presented to the man for interpretation: What could the coil be?

> That means that he has to subsume it, as to its typicality, under various typical prior experiences which constitute his actual stock of knowledge at hand. But not everything within the latter is used as a schema of interpretation.
>
> (Schutz 1970: 36)

Only those areas that are typically familiar to that which is perceived are used. These are 'relevant elements for his interpreting of the new set of perceptions' (Schutz 1970: 36). Interpretative relevance has two functions: it is not only important that an element of our stock of knowledge is relevant to the object, but also that only certain features of the perceived object are relevant to our interpretation. Thus the weight, length, size and possibly even the colour of the coil are not relevant to our interpretation of it when we are deciding whether it is a rope or a snake.

It is however important that a typical coil of rope and a typical curled-up snake are already established as types in the agent's stock of knowledge. The more familiar the man is with these types, the more clearly he can interpret the thematic object. If the interpretative reference points are inadequate, more and more interpretations of the same object are possible, without a particular interpretation being favoured. If one interpretation is to be selected, it may not be the correct one but at least it is the most probable. For this interpretative reference points must be as complete as possible. 'But in order to do this, I must must compare typical moments of the percepts with typical moments of my previous experiences of other typical rope piles or snakes' (Schutz 1970: 40).

This description of the interpretative process should not however lead to the misconception that 'interpretation' proceeds in a series of logical steps leading from premises to deductions. There are no isolated references points, nor are there isolated interpretative schemas. They are 'always interconnected and grouped together in systems' (Schutz 1970: 43). They are integrated into the stock of knowledge and linked to each other *via* systems of types which must be consistent if a probably correct interpretation is to be made. Yet, in line with my previous argument on falsification, every interpretation remains an *experiment*, however. It is always subject to 'verification and falsification by supervening interpretatively relevant material' (Schutz 1970: 45).

When the man prods the coil, or moves it in some other way, he does so because he is

> unable to come to an interpretative decision based upon the interpretatively relevant material at hand [...]. We have already stated that he does this because the diagnosis of the nature of this object is 'important' to him. 'Importance' is used here clearly related to the notion of relevance.
>
> (Schutz 1970: 45)

This has nothing to do with interpretation. It has to do with motivation, or *motivational relevance*. For the agent, the elements of a situation are always fixed in certain meaning-contexts which are also constituted against the background of the 'in-order-to' motive of the action, within the framework of stock of knowledge, 'because' motives. In every case, however, the result of the interpretation will determine the decision to act in a certain way. It is not only the meaning of the object that depends on the interpretation, but also the intended goals of the action. 'The satisfactory plausible degree of interpretation opens a relatively high subjective chance of meeting the situation effectively by appropriate countermeasures' (Schutz 1970, 46). Thus the project of the act is directly connected with the results of the interpretation. On the other hand, of course, thematic relevance and interpretative relevance are also dependent on motivational relevance. If the man in our example had not gone into the room *in order to* sleep, the coil would possibly have been neither thematically nor interpretatively relevant. In other words, 'the interest determines which elements of both the ontological structure of the pre-given world and the actual stock of knowledge are *relevant* for the individual to define his situation thinkingly, emotionally, to find his way in it, and to come to terms with it' (Schutz 1966: 123). The orientation of an act is thus in every case a selective activity of consciousness.

In the foregoing account, certain aspects of the action situation and of the stock of knowledge have been highlighted. They can be thematically, interpretatively or motivationally relevant. What remains to be clarified is the interdependence of the various relevance systems on each other and on the stock of knowledge. In the example quoted above, the motivational relevance led 'to the constitution of the "interest" situation, which in turn determines the system of topical [or thematic] relevance' (Schutz 1970: 66). Thematic relevance brings certain aspects of the field of vision from out of the background and into the thematic field of the agent's attention, 'thus determining the problems for thought and action for further investigation' (Schutz 1970: 66). In this way thematic relevance selects aspects from the given 'world' of action, which is otherwise taken for granted. Thematic relevance also determines to what level, and, to what extent, the agent has to gain the requisite knowledge to deal with the immediate situation. 'Thus, the system of interpretational relevance becomes established, and this leads to the determination of the typicality structure of our knowledge' (Schutz 1970: 66).

It could be mistakenly assumed here that the relationships between the relevance systems are chronologically linked. Schutz does not believe this to be true. In fact 'all three types (of relevance) are concretely experienced as inseparable, or at least as an undivided unit, and their dissection from experience into three types is the result of an analysis of their constitutive origin' (Schutz 1970: 66). All three relevance systems are, however, concerned with the field of consciousness, with the object of interest for the action in question. Thus any of the three relevance systems can become the chronological starting point for interpretation. During the interpretation, however, other aspects of the object may become thematically relevant and be reconstituted. Similarly, the acquiring of new themes can awaken new interests or establish new 'in-order-to' motives, and so on. At the same time, all three relevance systems are subject to constant modifications which are connected with the expansion of the stock of knowledge through new experience, or its 'contraction' through forgetfulness. Knowledge typifications consequently become more comprehensive or more limited. This, in turn, has an effect on the relevance systems, and on the adequacy of the frame of reference for the action's orientation.

Now that we have examined the various ways in which the acting subject relates to the frame of reference, we can clarify the ways in which a shared understanding of various goals is possible under these more precisely defined preconditions. As already stated, the approach to the problem of relevance has hitherto been concerned only with acts which relate to the material world. This approach gives the impression that 'an isolated individual experienced the world of nature disconnected from his fellowmen' (Schutz 1970: 73). Since Schutz wished to set up a sociological theory of action, it was imperative that he take account of the various relevance systems when analysing the problem of the orientation of acts relating to the social context.

The way in which Schutz sees an intersubjective understanding concerning action situations and orientations, in spite of the dynamic relationships between the various relevance systems, is revealed by the following observation. The knowledge with which we classify, interpret and incorporate the world into our various intentions and purposes 'is from the outset socialized knowledge, and thus too, are the systems of relevances' (Schutz 1970: 74). Unlike Parsons, however, Schutz's view of socialization is more complex. He continues his argument by emphasizing that the relevance systems are different for every individual and in every biographical situation. However, such divergence in the 'natural attitude' is not considered problematic 'for the time being'. Agents interacting with one another start out from the validity of another, different idealization.

The idealization (by which Schutz means the construction of ideal situations, similar to ideal types) of 'congruence of relevance systems' pertains to the fact that my biographical situation (and my corresponding relevance systems of classification, interpretation and motivation) cannot coincide completely with that of my part-ner. Nevertheless, in everyday life we start from the premise that the meaning ascriptions for objects and acts, which result from the relevance systems or are constituted by them, are adequate and congruent enough for communication, cooperation and colla-boration.[39] 'For the time being', i.e. as long as the acting subject's interpretations and knowledge survive testing by a partner without rejection, he or she can assume that the partner's perspective is sufficiently reciprocal for the immediate practical situation. 'I assume [. . .] that he, that we interpret the actually or potentially common objects, facts, and events in an "empirically identical" manner, i.e. sufficient for all practical purposes' (Schutz 1962: 316). Thus the 'general thesis of reciprocity of perspectives' explains the possibility of the mutual understanding of several agents, and thus the possibility of society itself.

The action situation

There are two main components in Schutz's action situation:

1 *The ontological structure of the pre-given world.* He explains this aspect as follows: 'To make a glass of sugared water [. . .] I must

wait until the sugar has dissolved' (1966, 122). That is, agents find that in their acts they are subject to certain (for instance, physical) conditions imposed upon them, which are outside their sphere of influence. These are the 'constraints' of an action met with in a certain situation.

2   *The biographical circumstances of the agent,* i.e. the nature of the agent's stock of knowledge at the time of the projection of the act and the action. 'The biographical state determines the spontaneous definition of the situation within the imposed ontological framework' (1966: 122).

In Chapter 3, I discussed the role of other agents in falsifying and testing knowledge about a situation. Where factors in the stocks of knowledge of several agents coincide, the situation (as a factor in the life-world) may be described as an intersubjectively shared one. It is a situation which is constituted and interpreted in the same way by all participants. If the life-world factors diverge, there must be a process of communication regarding the meaning of the facts relevant to the agents in that situation, until a consensus is reached, an intersubjective interpretation.

As far as the physical aspects of the situation are concerned, the agent refers to lived experiences in the 'life-form of the acting I', to bodies' experiences of 'becoming active'. Superimposed on these, however, are experiences in the life-forms of 'the Thou-oriented I' and 'the speaking I', the socially conveyed knowledge by means of which the ego can relate to the social aspects of the situation. The agent defines action situations within the experience horizons of the respective aspects: in the physical sphere bodily experiences, and in the social sphere the existing results of attempts to match personal constructions with those of the interaction partners. In fact the physical sphere, and Schutz's notion of space more generally, is critical to his theory of action. This will become clear in Chapter 7. It is enough at this point that, within the framework of situation definition, an 'order' is established by the agents/interacting subjects (in their subjective perspective), as a result of acts of interpretation. For Schutz, this is considered the most important precondition for social action and the social world in general.

SUMMARY

Our systematization and analysis of the various action-theory approaches has shown that Pareto's model to a large extent agrees with Popper's postulates, and that Schutz's model accords with the phenomenological foundations of the social sciences. The norm-oriented model of action sits between these two poles.[40] Pareto's model accords with Popper to the extent that both start from the premise that objective knowledge exists independently of the agent, and both see this as a criterion for the correctness of the agent's action. Both take the view that it is only very rarely that human acts are rational in the empirical sense. Further, both start from the premise that human beings act in accord with the deductive principle. Weber's model agrees with Popper's thesis to the extent that objective meaning exists independently of agents. It is not this context, however, which constitutes Weber's primary research interest, but the result of action orientation, where an agent combines objective meaning with subjective activity. In contrast to Popper, Weber takes the view that the primary aim of social research is to understand the endowment of subjective meaning, if the social sciences are to explain social processes properly. Rational choice theory attempts (tacitly) to reconcile the two approaches. The point of departure of rational choice theory is establishing subjective components. With these in mind, choice principles can be recommended on the basis that they approximate most closely to the objectively correct action.

Parsons' approach goes beyond the 'means–end problem', and places the norm-oriented components of acts at the centre of the debate. His overall position in relation to action is none the less unclear. On the one hand, his differentiation of the orientations frame of reference into personality, social and cultural systems seems to relate to the subjective aspect of goal orientation. But on the other, courses of action can only be assessed from an external objective perspective. In this context Parsons' approach lies midway between the objective perspective of Popper and the subjective perspective of Schutz.

Weber's postulate that social research should be carried out in the subjective perspective is treated much more radically by Schutz. However, Schutz does not start from the premise that the subjective consists merely of personal interpretations of objective social meaning. In his view, it is the other way round: 'objective' meaning results from intersubjective meaning-endowment alone. 'Meaning' is not 'attached' but constituted by the knowing and acting ego, on the basis of a stock of knowledge in a particular biographical situation.

We can now see why each of these different approaches can be relevant to empirical research, depending on the nature of that empirical research. The models of purposive-rational action, and to a certain extent also the model of norm-oriented action, are more precise representations of the categories for describing and explaining human activity within the framework of the objective perspective. They agree with the basic principles of Popper's procedure of situational analysis. If the meanings of the elements of the problem situation are unproblematical, if they lie within an intersubjective meaning context for the agents under investigation and the social scientist, the procedure of situational analysis is perfectly adequate. If the optimum choice of means constitutes the central aspect of the problem situation, where there are accepted goals, situational analysis can be carried out in purposive-rational action models. If social norms and cultural values constitute the central problem in the choice of goals or means, the corresponding analysis must refer to the norm-oriented model of action. If meanings are not perceived in the same intersubjective meaning context by the social scientist and the subjects under investigation, the relevant empirical research must be carried out within the framework of the subjective perspective. The action model of intersubjective understanding might then be applied.

We shall see that the same holds for research in social geography, exept for this: in social geography, it is crucial to examine action theory's reinterpretation of the processes of production and exploitation of spatial patterns in more detail. These need to be related to existing theories of space in geography, and it is to this I shall now turn.

Chapter 6

Geographical space and society

Traditionally, human geographers have analysed societies in terms of their so-called spatial character or in spatial categories. They have explained the social world in the context of spatial differentiation and tried to solve problems arising in this context (unequal opportunities in different regions, territorial conflicts, regional struggles, etc.). In this chapter, I shall argue that the spatial point of departure does not and cannot provide solutions to these problems. The action theory of social geography I am now introducing, a theory which emphasizes subjective agency, might do so. In this theory, of course, subjective agency has to be understood in its socio-cultural and physical–material frames of reference.

My argument here is *not* that 'action in space' should be the basis of research in this new social geography, as suggested by some geographers.[1] Since every action takes place 'in space',[2] this approach would contribute nothing to an explanation of different social systems from the 'spatial' point of view. I should also stress that in an action-oriented social geography, as in any other action-oriented social science, it is not 'space' which is the central unit of analysis, it is the 'action' and the 'act'. But this emphasis does not mean that the spatial dimensions of social reality are insignificant. I attempt to assess their importance for the performance of actions, but this does not contradict my devaluation of space in the geographer's conceptual hierarchy. Spatial relationships are only a precondition for a more comprehensive social theory. They can help us find a geographically appropriate way for differentiating the ontologically different areas of reference of human action. But the main tasks of geographical social research based on an active action theory are: (i) to understand and explain human actions;

and (ii) to clarify the relation of human actions to the social and physical worlds.

In the following section I shall develop these preliminary remarks. In later sections the sociological preconditions for the development of an action-oriented social geography will be investigated. We shall ask how far the models of purposive-rational and norm-oriented action and the model of intersubjective understanding take account of the physical–material or spatial aspects of actions. This question matters a lot, as geographers frequently maintain that sociological theories in general neglect the spatial dimension. In fact, the recent enthusiasm some geographers have expressed about Giddens' initiative in integrating spatial considerations into social theory (Giddens 1984; 1990) indicates how deeply the sociological neglect of 'space' has been felt. Geographers' interest in Foucault (1979) also indicates that there is strong feeling about re-examining spatial questions from a potentially more productive standpoint. But these expressions of interest, or attempts at applying Giddens and/or Foucault, reveal just how wide the gap is between the mechanistic concepts of space that have dominated geography, and explanations of human action. As I indicated at the outset, one of my aims in writing this book is to provide an account of subjective agency that will help bridge this gap.

Our first task is the differentiation of social, mental/ subjective areas and physical areas in a 'spatial' context. To do this well, we should try to locate the position of social meaning in the social world, a mental meaning in the subjective world, and a material object in the physical world. In terms of concrete actions, we can also distinguish which position the physical–material or biological vehicle (artifacts, the agent's body) takes up in the physical world as the 'carrier' of subjective meaning, and which position the expressed meaning assumes in the social and subjective worlds. Through a comparison of the various classification patterns, the connections and meanings of the various aspects might be more adequately understood than has thus far been the case in social research.[3]

Before we can go on, we need to examine the concept of space as geographers presently understand it. By analysing this concept, we should gain greater access to how the three worlds work in human action, and resolve certain paradoxes in connection with 'space' in general.

THE PROBLEM OF SPACE IN GEOGRAPHY

There is a sense in which Albert Einstein's (1954: xi–xii) formulations about space sit beside the geographical use of the concept. Einstein writes:

> In the attempt to achieve a conceptual formulation in the confusingly immense body of observational data, the scientist makes use of a whole arsenal of concepts which he imbibed practically with his mother's milk; and seldom if ever is he aware of the eternally problematic character of his concepts. He uses this conceptual material, or, speaking more exactly, these conceptual tools of thought, as something obviously, immutably given; something having an objective value of truth which is hardly ever, and in any case not seriously, to be doubted. [. . .] And yet in the interest of science it is necessary over and over again to engage in the critique of these fundamental concepts [and 'space' is one of them] in order that we may not unconsciously be ruled by them. This becomes evident especially in the situations involving development of ideas which the consistent use of the traditional fundamental concepts lead us to paradoxes difficult to resolve.

But there is one important difference between Einstein's comments and the present situation in geographical research. While the majority of geographers are indeed unwittingly governed by a particular concept of space, and while this often leads to paradoxical arguments, they are for the most part unaware of the problematical nature of the fundamental concept upon which their research is based. Geographers in general do not think that 'space' only refers to a concept which enables us to 'order the immense mass of experience'.[4]

A first, perhaps a decisive, step towards clarifying the use of 'space' in geography would be a consistent definition of 'space'. The second step would be, as in any definition of a concept, to distinguish between that which is signified and the sign or word which signifies. As Saussure has taught us, the object of experience and the combination of letters used to label it can never be identical, and consequently cannot be equated with each other (Saussure 1960). When they are equated, concepts are made concrete in an untenable way.

An important feature of concepts of space is that they are neither prescriptive nor empirically descriptive. They help us to order and structure our experiences, but they do not relate to one given object or fact. They relate only to a formal aspect of all objects, which can be abstracted from their other features.[5] In other words, if we have defined an object in terms of a concept, 'space', we have not established any of its other features. It is therefore pointless to deduce anything about an object on the basis of its spatial position alone. Spatial location always and only expresses a formal aspect of an object.

But while every object can be defined or located spatially, regardless of its content, objects which are otherwise identical can be distinguished by their spatial location. The categories used to locate an object in this way depend on its prevailing definition of the concept of space. Every definition establishes particular dimensions by which an object is ordered and located. Yet because a spatial frame of reference is only one form of classification, familiar geographical terms such as 'spatial problems' are misleading. There can in fact only be 'spatial problems' when it comes to defining spatial concepts. Rather than talk of spatial problems it would be more accurate, and more meaningful, to talk of problems with the way material objects on the earth's surface are arranged. And as a rule, once we focus on such arrangements, we shall have to talk of the actions and circumstances that lead to those arrangements. And these arrangements are, at a given moment, only problematic in relation to aims and goals or to other aspects of a specific action. What geographers term 'spatial problems' are in fact problems relating to action.

It should be added that one single concept of space cannot be suitable for all kinds of facts [*Sachverhalte*]. Different spatial frames of reference must be defined for ontologically different factors, if they are to make meaningful contributions. If this point is overlooked, the inevitable result will be reductionism and empirically unsound conclusions.

As I noted in Chapter 1, the geographical use of the term 'space' tends to treat it not as a concept but as a *thing*. 'Space' is frequently fetishized and endowed with the capacity to cause particular events. This fetishism might justify the dominant definitions of geography in general and human geography as a spatial science in particular. In both disciplines, the formal condition of events is held to be its cause. Just as 'language expresses thought

processes which occur *in* words but not *by means of* words', as Georg
Simmel (1903: 28) wrote, so events can only be spatially located.[6]
They cannot be caused by 'space'.

The main problems connected with geographical spatial understanding are: (i) either the whole physical world, or the part
experienced and perceived from a certain standpoint, is viewed as
concretized space (the landscape); (ii) a formal property of an
object is taken as its cause; (iii) socio-cultural phenomena are
explained in spatial terms.

If 'action' rather than 'space' becomes the central theoretical
concept of social geography, the spatial *arrangement* of objects
becomes relevant as a necessary condition and consequence of
human action rather than a cause. Actions have goal-specific
orientations in relation to locatable objects. Locatable objects, of
themselves, do not *cause* acts, although of course they frame them.

Despite his later spatial reductionism, Bartels once gave a useful
definition of the significance of spatially locatable elements for
action. Most human actions 'have to do with material things, with
exclusively spatial considerations' (Bartels 1974: 19), which can
also be of great importance for the social context of action. At the
same time, in many of their social interactions agents have to
overcome the distances between them. And the decision processes
of most agents searching for the best positions for their actions
have to take spatial factors into account. The relation touched on
here between phenomena in the social world and those in the
physical–material world, between their respective patterns of
arrangement and mutual connections in courses of action, should
be seen as the central theme in an action-oriented social
geography. A social geography of this kind should be able to
integrate patterns of spatial arrangement, their effects on human
actions, and the latter's interventions in the physical world
systematically and consistently into its theories.

If we are to understand the significance for social processes of
the spatial arrangement of physical–material conditions and
means of action, the actions themselves must also be understood
and explained from both a social and a subjective point of view. At
the same time we need adequate frames of reference for every area
to which actions relate. This thesis is based on the idea that agents
themselves are using different frames of reference for their orientation in the physical, subjective and social worlds. They need to do
so to determine their own positions in this respect, and the

positions of that which they wish to achieve. It is only in this way that agents can establish the 'path' down which they are to proceed. Social researchers also need a reference pattern of orientation, if they are to provide empirically valid explanations. In the following discussion, I term all these reference patterns 'spatial frames of reference'.

The orientations which the agent adopts in concrete situations of action in relation to the physical, social or subjective worlds have to be systematically and meaning-adequately reconstructed. To this end, spatial frames of reference must be developed whose definition establishes particular characteristic dimensions which coincide with the ontology of the (physical, social or mental/subjective) object to be located. An appropriate spatial frame of reference should thus be made available for every ontologically different world to which human actions may relate. This is a critical theoretical precondition for analysing the relationship between the socio-cultural, subjective and physical–material dimensions in human actions. The differentiation of the concept of action in turn depends on effective schemes of reference for the locating and structuring of the situational elements relevant to action. As we saw in earlier chapters on epistemology, these elements belong to the three ontologically different areas. In other words, for each ontologically different area of reference for human action, specific concepts of space must be found for human geography's analysis of society.

SPATIAL FRAMES OF REFERENCE IN THE OBJECTIVE PERSPECTIVE

We have seen that Popper's precepts for the social sciences start out from this premise: human actions can be explained through reference to the conditions of the action situation.[7] He ascribes a subordinate role to the subjective sphere (world 2) in social research, because this sphere is in any case directed by the social and other components of world 3. It is therefore primarily the areas of reference of the physical and social worlds for human actions which are important for social research in any social geography which follows Popper's procedure. For a truly Popperian geographer the question is: Which reference schemes enable us to structure the two worlds from an objective standpoint, thus clarifying spatial locations and orientations?

The spatial frame of reference for the physical world

Let us consider some Popperian examples. Remember that Popper defines world 1 objects as physical–biological–material. The agent's body also belongs in this category. One scheme of reference, in natural sciences and geometry, for locating objects in the physical world is described as a physical–geometrical spatial concept (Sorokin 1964: 97ff.) or as a mechanical–Euclidean spatial concept (Jammer 1954). In this concept two different objects in the physical world cannot occupy the same location at the same time (mutual exclusiveness). More specifically, the categories used in this concept in describing the positions and movements of material objects are 'length', 'width' and 'height', and a 'point of origin' where these three axes meet. When the point of origin has been chosen, any point within the three dimensions can be precisely determined. In describing the movement of objects from one position to another, one has to deploy other categories: namely, the direction of the movement (vector) and the object of the movement. For the description of distances between the positions of material objects, or of the distance covered by means of movement, there must also be a unit of measure (metre, yard, etc.).

On the basis of these categories, both the positions and the movements of objects can be objectively determined. We can construct an isotropic and homogeneous scheme of categories, whose point of origin can in principle be placed anywhere at all, depending on the aim of the investigation.

In the geographical adoption of the mechanical–Euclidean concept, not three, but two, dimensions (longitude and latitude) are used in spatial definitions.[8] By means of geographical lines of latitude and longitude – relating to Greenwich or some other point of origin – we can stipulate points in space in the global context. The direction of movements can be established in relation to the points of the compass. The distances between positions, and the distances covered/coverable through movement, can be measured in terms of units of measurement which have been generally agreed upon.

But apart from its physical–geographical application,[9] this concept of space is also used in human geography in investigating regularities of distribution. How does Popper's theory fit in here?

Although Popper ascribes no importance to agents' specific goals in the explanation of actions, from the physical–spatial point of view, goals are also important for the objective perspective. In order to attain a physical–spatial goal (e.g. a meeting point in town), the agent can, after establishing a personal position in the objective 'two-dimensional frame of reference of measurement', avail himself or herself of the categories of the mechanical–Euclidean spatial concept on which maps are based. In other words, agents can look up maps, etc., and work out where they are in relation to where they want to go. Obviously this is useful. But it only tells us about the material location of objects and bodies. It tells us nothing about socio-cultural meanings.

Bollnow (1980: 28ff.) and Lefebvre (1981: 31ff.; 265ff.) object to the mechanical concept of space. It is a concept produced by capitalism, and is 'inhuman'. It disregards socio-cultural meanings and context, and should therefore be rejected. The problem with their position is that it rejects spatial considerations *tout court.* Evidently these considerations have a physical–material use. The real problem is that geographers attempt to use a spatio-mechanical concept in locating non-material objects, such as socio-cultural (and subjective) meaning, for which it is useless. At the same time, the attempt to relate the socio-cultural context to spatiality should not be abandoned. Rather, the conditions of that relation need to be rethought.

A spatial frame of reference for the social world

Jarvie (1972: 161ff.) states the view that agents orient themselves in the social world by means of a 'mental map'. On this they mark out the positions and paths by which they can achieve their goals in the social world (e.g. changing a particular law perceived as unjust). He maintains that the central elements of this world are institutions such as 'the market', 'unions', 'the judiciary', etc., which occupy a certain position at a particular time. For the time being we can treat Jarvie's view as a more precise version of Popper's idea (1979: 74) that discoveries can be made in world 3 in the same way as they are in the physical world. Agents can orient themselves successfully if their social 'mental maps' coincide sufficiently with the empirical order of social phenomena. A researcher, too, can adequately explain an action through Popperian situational analysis if his or her mental map coincides with that of the agent. This spatial metaphor has so much appeal that it warrants more investigation.

Sorokin (1964: 103ff.) lists a series of attempts to construct a spatial model for the social world based on the physical–geometrical one. Socio-cultural phenomena always make their appearance somewhere: they always occupy a *position* in the social context and relate situationally to other phenomena. This led Sorokin, together with Carey, Pareto, Barcelo, Haret and others, to the conclusion that some spatial concept must be devised to describe and structure socio-cultural phenomena. Still, socio-cultural phenomena as pure meanings are not material in character, having neither length, size nor volume. Sorokin at least recognizes that it is therefore inappropriate to include them in the physical world. 'Pure meanings are entirely outside any kind of geometrical space, be it the three-dimensional space of classical mechanics, or Lobatchevsky's, Riemann's, or any other space of mechanics, geometry, and topology' (Sorokin 1964: 114).

Positions coinciding with the meanings of actions or subjective agency in the social world can be very close to each other, even though their vehicles (material artifacts, the agent's body) are far apart in the physical world. Movements from one spatial position to another can be measured physically, whereas movements from one position to another in the social world are not *a priori* linked to a change in the physical world. The less weight we give to this distinction, the less capable we are of understanding the social relations of physically separate phenomena, or that physically close phenomena may have no social relation at all.

There is another way in which the social and physical worlds are independent of each other. The same material objects can symbolize different social meanings, and the same meaning can be represented by different objects. *A priori* the location of an object in the physical world tells us nothing about its position in the social world, and *vice versa*: the physical–biological features of an object and its movements in the physical world do not adequately determine its socio-cultural meaning. This independence means that the two worlds must be separately analysed. A physical–biological object acquires various qualities of meaning for the agent, according to the particular socio-cultural context into which it is integrated through the performance of actions.

It should be evident by now that traditional spatial categories, despite their great success in establishing positions and orientations in the physical world, are totally unsuitable for 'structuring' the social world. The empirical relevance of research on action in the spatial dimension, and of all cultural–geographical attempts to explain cultural and social processes in a one-sided spatial way, is thus fundamentally in question. So we have a negative answer to Bartels' question (1974: 18) whether 'spatial measurement can be used at all' for the social world of human action, 'a primarily immaterial world, consisting of decision situations and value relations'. But we are left with the question: How can a spatial frame of reference relate to the social world and a successful orientation within it?

To establish this relation, the following requirements need to be satisfied. Social scientists must be aware of the coordinates of the social world. They can then locate the meanings or goals of actions, and explain them in an empirically valid way. In addition, researchers need to know how meaning, interest and expectation are positioned in relation to the action under investigation. They should also be aware of the distribution and stratification of particular social positions from which actions are undertaken or to which actions relate. It should then be possible successfully to reconstruct the 'paths' leading to goal-attainment in the social world. Social scientists should also be able to determine the position of a particular socio-cultural phenomenon in relation to other ones.

On the first point: the position of a meaning is determined when we are able to locate its social context. Accordingly, we need a way of connecting this social 'location' to a spatial frame of reference.

An important step in this direction was undertaken by Sorokin. His idea can be understood as a more complex version of Popper's world 3. Sorokin thinks that meaning exists 'as such', and not only in a subjectively constituted context. I shall now examine his argument as a possible point of departure for the problem at hand.

For Sorokin (1964: 122ff.), every phenomenon or element of the socio-cultural world (which for the sake of brevity I shall refer to as the 'social world') has three components: (i) the objective, pure meaning as such; (ii) a medium (vehicle) in which this meaning is expressed; and (iii) an agent. Sorokin wants a spatial frame of reference which could enable each of these three aspects to be positioned. In terms of action theory, we can reformulate this approach in this way. For the subject, objective meanings are (following Weber's idea) subjectively bound up with actions, and manifest in the course and consequence of action, which are either material or non-material in nature. We therefore need to concern ourselves only with meanings and their location. The position of the agent can be deduced from the typical expectations of the actions of a subject. We can establish the position of relevant media from the meaning attached by agents to their production activities, or attached to them at the time by the user. My interpretation is consistent with Sorokin's idea, in that he himself states that the location dimensions for all three aspects are basically the same. In other words, the positions adopted by the performer of an action and those of its media or vehicles are in principle defined by a meaning system, and can be deduced from it.

Sorokin's spatial frame of reference for the social world comprises the following dimensions: 'religion', 'science', 'language', 'art' and 'ethics'. Sorokin divides each of these primary dimensions into more detailed subdivisions, attempting to establish the locations of the meaning of actions more precisely.

- *'religion'*: the different systems of religion in existence now, 'Christianity', 'Taoism', 'Hinduism', 'Buddhism', 'Islam', 'Judaism', etc., and further subgroups within these such as 'Catholic', 'Protestant', etc., in 'Christianity'.
- *'science'*: scientific disciplines such as 'mathematics', 'physics', 'chemistry', 'biology', 'sociology', etc., and their institutions, such as 'primary school', 'grammar school', 'technical college', 'university', etc.
- *'language'*: the various language systems such as the Indo-European, Latin, Slavic, etc., and their subdivisions and hybrid forms such as 'German', 'English', 'Danish', etc.
- *'art'*: the various forms (music, literature, . . .) and styles (Baroque, Dadaism, . . .).
- *'ethics'*: the various ethical codes, corresponding legal systems and their further subdivisions (civil law, constitutional law, criminal law, etc.).

Sorokin also isolates, though without giving reasons, two 'in-between' categories:

- *'economy'*: the various forms of economy within which trade is carried on, such as 'market economy' or 'planned economy', etc. (part of 'ethics').
- *'politics'*: the various political systems.

For Sorokin, the meaning of an action or its consequence is located adequately if we are able to establish its position in one of the above-mentioned dimensions. Yet if we compare this schema with the physical–geometrical concept of space, we see that an important feature is missing: namely, the 'point of origin' as the central category of orientation. Sorokin does not specify this, nor accordingly can he isolate a unit of measurement for determining distance between positions of pure meaning in the social world. For him it is sufficient that two meanings are close to each other if they occupy adjacent positions in the same dimension, and far apart if they occupy positions distant from each other in the same or even in another dimension. If the meaning of action A is religious (e.g. Baptist), and that of action B economic (e.g. an American oil company), they accordingly occupy positions very far apart in the world of (social) meanings. This is still the case even if the bodies of the agents performing these actions are very close to each other in the physical–material world.

Overall however, Sorokin's proposals, while unconvincing in many respects, are a step forward compared to the rudimentary ideas of Popper and Jarvie on the same subject.

Another proposal is put forward by Parsons and Bales (1953: 63ff.). On the one hand, they want to reconstruct the scheme of reference which an agent uses for orientation in the social world. On the other, they wish to investigate the processes taking place in that world in relation to spatial categories.[10] In their view (1953: 88) the advantage of this approach is that – if the idea succeeds – a generalized scheme of description and explanation of actions would be available which could be applied both to the social world ('action space') and the physical world ('behavioural space'). They hope that, on this basis, human actions may be understood in relation to both the social and the physical world. They think that a spatial frame of reference for the social world should be constructed after the manner of a spatial model for the physical world. Parsons and Bales suggest that the spatial frame for the social world should be understood in Euclidean spatial terms, even though their social model is conceived in four and not three dimensions and does not aim to locate material objects.[11] The spatial frame for the social world is Euclidean in the sense that 'it is "rectilinear", that there is continuous linear variation along each of the dimensions, and that time enters into the analysis of process in essentially the same way that it does in classical mechanics' (1953: 85). Without expanding on this remarkable proposition, Parsons and Bales go straight on to formulate further requirements. The spatial frame of reference for the social world will make possible the location of the 'particles' of that world and a clear description of their movements within the action process, 'in terms of the four coordinates of the space' (1953: 86). As for the 'point of origin' they state (1953: 86): we must

> be able to describe the location at an initial time (t_1) and a difference of location at a subsequent time (t_2). Each location must be described in terms of *four logically independent statements of fact*, one for each of the four coordinates of the space, hence change of location must be definable as change relative to each of the four coordinates.

So what remain to be defined more precisely are the dimensions, the 'point of origin', the uniform standard of measurement used to determine movements, and the 'particles' which have to be located in relation to the movements and described in relation to their change of position during action orientation.

On the 'point of origin' to which the four dimensions must relate, Parsons and Bales do acknowledge that it is doubtful whether one can proceed as arbitrarily for 'action space' as for three-dimensional Euclidean space: 'The problem seems to be connected with the fact that in some sense there are "boundaries" to the space of theory of action, to which there are no close analogies in the space of classical mechanics' (Parsons and Bales 1953: 91). They reach the following conclusion (1953: 95): 'the point of origin cannot be arbitrary [. . .], [it] must be relative to the particular system which is being analysed'. This means in effect that establishing the point of origin depends on whether the analysis relates to the 'personality system' or to the 'socio-cultural system'.

As for the uniform standard of measurement necessary to determine the movements of 'particles' in relation to the point of origin and its relevant dimensions, they maintain (1953, 90) that there can be no absolute, but only relative, standards of measurement. It is enough to determine whether the respective particles have approached the reference point in these dimensions, or whether they have moved away from it.

In his later works Parsons did not clarify or elaborate on this first spatial outline for the social world, and much of it remains obscure. In particular, the 'point of origin' remains a mystery, especially if we assume with Parsons that in every act there is interpenetration of personality, social and cultural systems. Above all, Parsons and Bales fail to establish a connection between their spatial concept of the social world and a reference pattern for the description of activity in the physical world. As Konau (1977: 196) succinctly remarks, Parsons' linking of the spatial concepts of the social and the physical worlds is more a statement of faith than a solution to the spatial problem.[12] In sum: Parsons *et al.* (1953: 200) stress that we must systematize orientation processes in the social world just as we are able to in the physical–material world. Their statement: 'We begin to feel the need of a representational model more in the nature of a "space" in terms of which all the relations can be represented simultaneously' (1953: 200) points in the wrong direction, however. For the simultaneous location of elements from the social and the physical worlds in *one* scheme of reference always implies an untenable reduction which leads to statements which have no empirical value. My theory, that a specific scheme of reference is necessary for every ontologically different area of human action, is reinforced accordingly.

The most recent contribution to the discussion of a spatial frame of reference for the social world is that of Bourdieu (1984a and b, 1985, 1991a, b). His point of departure is the observation that the Marxist conceptualization of the social world is based on an economistic one-dimensional form of social space. By this he means that the Marxist theory is reductionist: the multidimensional social world becomes a one-dimensional social space (Bourdieu 1984a: 9):

> the relations of productions became the one and only category for the ordering of the social, cultural and political fields: the Marxist theory of category for the ordering of the social, cultural and political fields: the Marxist theory of classes [. . .] is the result of the reduction of the social world only to the field of economics. This theory defines a social position by referring only to the position of an agent in the relations of economic production and is ignoring at the same time the occupied positions in the other fields and sub-fields. By this the Marxist theory is constructing a one-dimensional social world built around two blocks (owners and non-owners of means of economic production).[13]

In addition an adequate concept of social space has (i) to overcome the substantialism of the Marxist view and to be replaced by a relational interpretation; (ii) scientific constructions should not reify 'classes' and take them for 'real'; and (iii) a false objectivism in the representation of the social field has to be abolished (see Bourdieu 1984a: 3).

So instead of a one-dimensional concept we need a multi-dimensional social space. For Bourdieu (1985: 9) this space is produced by different principles of distribution. This means that the positions of *agents* are defined by a relative position to each other and every agent occupies in reality only one position in the social space. The principles of distributions are seen by Bourdieu as the originating force for the structuring of the social space as a field of force. These principles are 'power' and 'capital'. But, in contradiction to the Marxist view, Bourdieu distinguishes different types of 'power' and 'capital' and not just one.

In fact he distinguishes four different types of capital and four related fields of the social world. They are: (i) the economical field: economical or material capital; (ii) the cultural field: cultural capital (one sub-field of the cultural field is the field of literature (Bourdieu 1991a: 11) divided into further sub-fields as journalism, theatre, cabaret, novels, poems, etc. (Bourdieu 1991a: 31)); (iii) the social field: social capital; and (iv) the symbolic field: symbolic capital (see also Bourdieu 1991a: 4). The last one he sees as the perceived and legitimately accepted form of the three other forms of capital, normally labelled as prestige, reputation, etc. Every type of capital determines the possible type of benefit in each field. More implicitly, Bourdieu (1984a: 10) is sometimes referring to a fifth and sixth field – the field of power and the field of politics – but they are not elaborated in detail.

Every real position of each agent in the social space is therefore the sum of his theoretically constructed positions in each of these fields and every specific position is defined by the means of power in the respective field. That is: every position of an agent is defined by the coordinates of the multi-dimensional social space. His or her position depends on (i) the total amount of capital at his or her disposal and (ii) the composition of this capital, the proportional importance of each of the four capitals in the total amount of capital. And this position determines the relations of power in the field of force, in the social space and the respective chances of benefit.

On the bases of these positions, social scientists are able now to build up social classes in the logical sense of the term. This means:

every social class contains agents with similar positions in the social space and the corresponding dispositions and interests. The position determines therefore the dispositions, interests and careers of the agents, their praxis and ideological–political positions. For these reasons Bourdieu comes to the conclusion: 'These classes [. . .] enable the explication and the prognosis of the praxis and the attributes of the localized facts: and besides other things the organization of groups.'[14] But these 'classes' are not real, they 'exist only on paper'.[15] They express a space of relations and 'this space is as real as the geographical space,[16] even if 'these two spaces never coincide totally: even most of the differences attributed to effects of the geographical space, for instance the opposition of centre and periphery, are the expression of distances in the social space, namely the uneven distribution of the different types of capital in the geographical space'.[17] But there are other similarities: the change of positions 'costs in both spaces labour, performance and time'[18] and the distances are measured in both cases in time. The probability of mobilizing an organized social movement is 'inverse proportional to the distances in this [social] space'[19]

Political struggles in the interpretation of Bourdieu are nothing other than struggles to get a 'better' position in the social space, seen as a field of force, expressed by the relations between positions: 'the field of power is a space of the relations of forces of different agents or institutions [. . .] to occupy the dominant positions in the different fields (economical or cultural)'.[20] Or more generally: 'Social space' means that we cannot bring together any bodies occupying totally different positions. We have to respect all the distinct characteristics of agents in the economical and cultural field and the positions occupied by agents. This proposition is in many respect problematic and needs therefore some further elaboration. I shall now discuss some of the critical elements in relation to my theory of space and to action theory in general.

The first point to mention is that Bourdieu tries to localize agents and not socio-cultural meanings with his concept of social space. This is the consequence of the central role of the concept of 'habitus' in his theory of structuration (Bourdieu 1991a: 6): 'to analyse the "habitus" of the occupiers of these positions'.[21] Therefore this internally consistent with his theory. But, at the same time, it leads to a number of contradictions in terms of an adequate concept of social space.

If we take into account the agent's body we realize that Bourdieu's strategy must fail in many respects. He cannot localize bodies in his social space and in addition it reduces the potential variety of action produced by an agent to one type of relevant action. At the same time he is not able to make a clear distinction between the spatial concept of the physical and the social worlds. This is expressed in the statement where he mentions 'the uneven distribution of different capitals in the geographical space'. In fact, they cannot 'have' a position in the physical world, because they have a symbolic and not a material status, as Bourdieu points out quite correctly elsewhere. Instead of agents Bourdieu should try to localize meanings of *action* in the social space, which is to say the dominant meanings in economical, political, social and cultural actions in the respective fields or sub-regions of the social space. In fact Bourdieu contradicts himself. When he makes it clear that the symbolic field represents the perceived and accepted form of the three other fields, this would imply that agents have only symbolic features. But then it becomes very difficult to accept them at the same time as agents moving 'in' the geographical and social space at the same time. The bodies of agents have therefore to be localized in the spatial frame of reference of the physical world. What Bourdieu calls 'habitus' is part of the mental world and should therefore not be localized in the social world or the social space.

The statement that movements *in* the geographical and *in* the social space are both 'real', can be accepted, but in another sense, from that which Bourdieu claims. In my view, both are 'real' in the following way: moving in a spatial sense is 'only' an event under a certain description in respect to a certain concept or framework. Moving 'in' geographical space expresses the motion of material facts, moving 'in' social space indicates a 'change' of meaning contents in the framework of the immaterial social world. So both are only real under a certain description. But they never coincide. If they did, any change of a position in the physical world would imply a change of the social position and vice versa in a deterministic way. This is obviously not the case, even if very often certain positions of the physical 'earth-space' get – as an outcome of acts of *interpretation* – a certain economical, social or cultural signi- fication. But this signification is not a property of a position in itself, but a produced property in the framework of acts of interpretation.

The second point refers to the delimitation and the relations of the different fields. Bourdieu does not make clear why he is referring in two different ways to social space. On the one hand he uses it as the concept for the totality of the social world (1991a: 11) and on the other he refers to it as one field of it (1984a: 3; 1985: 10). If this is a problem of clarity, the other aspect is more substantial: it concerns the hierarchy of economical, cultural, social and symbolic fields and their interrelations (see Bourdieu 1984a: 10).

As indicated, Bourdieu's critique of the Marxist class concept focuses on the one-dimensionality and the pre-eminence of economic categories for the ordering of the social world. I agree with the first aspect of his critique and I consider it a very important point. Also I agree with the second point of his critique but Bourdieu himself cannot satisfactorily eliminate its weakness. First, he puts the field of economics at the top of his hierarchy and, second, this results in structuring the cultural and social fields in terms of economic categories (capital, mode of production, etc.). If these fields and the subsequent patterns of positions are the outcome of different principles of distributions it becomes questionable whether we can use for all expressions referring to *one* of these fields only. The result is a silent economization of the other realms of social reality. On the other side, the hierarchy, with the economical field at the top, remains within the Marxist world view. It is also questionable whether a fixed hierarchy of different fields is valuable for all types of social actions, even if the economical plays a central role in late-modern societies. It seems to me that any idea of hierarchy is rather an empirical question than a question of a theoretical decision a *priori*.

A further important point of critique concerns Bourdieu's understanding of 'space' and the role of 'space' for the ordering and orientation in the social world. He states, quite correctly, that we should not fall into the traps of substantialism and reification. But he cannot avoid them in his explicit and implicit definitions of 'space' and 'position'. First of all, Bourdieu often uses the expression of '*in* space'. That would presuppose a concrete 'space', a container, in which we can put the elements of location. If this is not accurate for the physical world, it is totally misleading for the social world. In fact this expression indicates that Bourdieu does not see 'space' as a concept for ordering and orientation.

The other point, concerning the status of 'field' and 'position', is even more important. Bourdieu (1984a: 10) allocates them a causal force. He describes the relations of the different fields and positions as a causal relation 'and the form of the causal determination is defined by the structured relations'[22] and the principles of distribution. It is the 'structure of the political field, i.e. the objective relations of the positions, which are determining the political acts'.[23] This interpretation of social positions comes very close to the structuralist point of view expressed by Althusser and Balibar (1970: 180): 'The structure of relations of productions determines the places and functions occupied and adopted by agents of production, who are anything more than the [. . .] supports (Träger) of these functions.' They also think that the position has a social quality and a social force in itself. But 'positions' cannot act or be an immediate causal factor of determination. The position cannot be an agent in itself, it can only be a representation of an attribute of a given fact, the description in respect to attributes of other given facts. Positions do not act; they are the outcome of definitions of certain circumstances for actions undertaken by agents at an earlier time. In that form they become relevant as supposed attitudes towards them, and expectations of them held by the acting agent. In this form *the expectations* linked with the (institutional) social position are of great (enabling and constraining) significance for actions. 'Positions' become only relevant in acts of interpretation, but not in themselves. In this respect, Bourdieu's idea of 'field of force' constituted by the different position is at the same time also misleading. Because the different fields are the product of a certain categorization in a certain respect, they cannot have any immediate force in themselves. That they become socially relevant through interpretation in courses of action is something other than attributing to them a force in themselves.

By neglecting this difference Bourdieu's approach comes very close to that of the spatial science of the social. If positions do not have an immediate force or power, it becomes meaningless to qualify the social space as a field of forces. If we ignore this we have to accept all the shortcomings of the spatial approach in the tradition of geography: not for reasons of vulgar materialism but for reasons of inadequate theoretization of social space.

Both the spatial science of the physical and the social world would be valuable only (i) if positions and distances really had a causal force and (ii) if all the agents' intentions could be oriented to the production of a certain space or to obtaining the causal force of a certain position. For these requirements it is difficult to find empirical evidence.

Instead of this we should understand concepts of space as means of describing and ordering facts without any force in themselves. We should bear in mind a clear distinction between the qualities of the localized fact and the constructed (theoretical) qualities of spaces and positions. They produce nothing. They are conceptional tools but not some kind of magic or mysterious agents behind social agency and/or action.

A spatial frame of reference for the mental world

As already mentioned, Popper sees the subject's mental world as 'merely' mediating between world 3 and world 1. It is primarily the objective content of world 3 which determines references to world 1. In the context of the spatial problem this means that no separate spatial frame of reference is necessary for the mental world. This world comprises the subjective representation of facts and states in the social and physical worlds, as well as their relation to the particular dimensions of corresponding spatial concepts in the objective sense. On the other hand, 'subjective space' should not be confused with the subjective representation of objective spatial dimensions.

Yet, in the 'mental map' of behaviourist geography it is. The 'mental map' is concerned not with subjective spaces but with the subjective representation of the spatial arrangements of material objects, based on spatial concepts of the physical world. Such studies could be useful for an action-oriented social geography, but only if they are placed in a different epistemological context. They should not be taken as the starting point or 'stimulus' of an activity, but as the goal-specific, subjective definition of physical space. They should only be interpreted in terms of orientation of bodily movements in the physical world. Similarly, the social 'mental maps' touched on by Jarvie should not be read as an account of subjective spaces. They are only mental representations coinciding to a greater or lesser extent with 'objective' facts and expectations.

Looking at Popper's argumentation as a whole, we can formulate the theory that within the framework of the objective perspective, an orientation in the context of both the physical and the social worlds will be successful if subjective representations coincide with the objective ordering of them. However, a particular problem becomes apparent here in the application of Popper's theory of knowledge to empirical social processes. Concrete forms of human association [*Vergesellschaftung*] do not take place solely in world 3, or world 1 or world 2. For the most part they evolve through a certain mediation between the physical and the social worlds. Here we touch on the aspect of territorial social systems, which cannot, on the formal level, be systematized using either physical or social spatial frames of reference alone. We should therefore pay particular attention to the mediation between the social and physical areas of human existence. Before turning to this, I shall first examine how the phenomenological approach bears on the spatial problem. I shall consider whether, in courses of action, there are any points of contact between the social world, as I have described it, and the physical world.

SPATIAL FRAMES OF REFERENCE IN THE PHENOMENOLOGICAL PERSPECTIVE

In Chapter 3, I showed that Schutz, too, has a three-world theory, although he did not define it as such. In his theory, both the physical and the social worlds exist independently of the knowing subject. However, according to his theory, meanings pertinent to either are always constituted first in the subjective world. Thus I shall concentrate first on the subjective world, and the corresponding scheme of reference for location, before going on to consider that of the physical and social worlds. I shall argue that Schutz does indeed link the three worlds by means of spatial frames of reference appropriate to each. In his theory intersubjective understanding and communication are only possible if subjects succeed in ordering their respective orientations to these three worlds in space and time.

A spatial frame of reference for the subjective world

The subjective world is constituted in the life-forms of pure and memory-endowed duration. As I discussed in Chapter 3, for Schutz

(1982: 48), 'to ask about the meaning of an experience means to look for the *place* of an experience-having past in the flow of pure duration'. The structuring of the subjective world by means of an adequate spatial frame of reference should accordingly start out from this basis: the sequence of sedimentation of knowledge at hand in the life-form of pure duration is of major importance. Although such a sequence corresponds primarily to a temporal structure, in action orientation it also becomes spatial. The question *when?* corresponds in the life-form of memory-endowed duration to the question *where?* in the process of the sedimentation of the stock of knowledge. This process of sequencing is, however, for Schutz not unstructured. It is specifically ordered according to types. If we understand 'types' as categories for ordering experience, on the basis of which agents orientate themselves within memory-endowed duration, we can say that 'types' structure this area of subjective experience. If the search for the relevant knowledge is adequate to the type of knowledge required in a particular situation, the agent has oriented himself or herself successfully in the corresponding space. The success of the action also depends, however, on the social intersubjective validity of types, as well as on their truth in relation to physical components.

To the question: by which main dimensions are the more general and complex types ordered? Schutz's answers (1962: 7ff.) are only vague. He states that the agent starts out from social types when he or she is constituting meaning. This might imply that we should structure the sphere of subjective experience in the social world according to Sorokin's or similar dimensions. But if we do, we should take these dimensions not as *a priori* objective criteria, but as the agent's intersubjectively valid criteria. The same applies to the concepts of 'nearness' and 'farness'. The same would have to hold for meanings relating to the physical world.

In addition, Schutz's (1970) statements concerning the problem of relevance help us answer the question of the 'point of origin'.[24] During the performance of action, 'interpretational', 'thematic' and 'motivational relevance' become the points of origin of orientation in the various dimensions of meaning, and the areas of meaning structured according to type. In contrast to the physical–spatial dimension, however, the point of origin of orientation should not be seen as fixed. Like the stream of inner duration, this 'point' is constantly changing in accordance with the relevant requirements for orientation at the time. This constant interplay, the prevalent relevance of interpretative, thematic or motivational aspects, is transposed to specific meanings in this world of subjective experience. And, as I have stressed throughout, the world of the agent's subjective experience occupies a central position in Schutz's thoughts.

A spatial frame of reference for the physical world

Space in the physical world is constituted *via* the experience of the subject's own body through the conscious self in movement. The agent thus experiences the physical world and represents its spatial dimensions *from the perspective of his or her own body*. Similarly, the materiality of the physical world is experienced through direct bodily contact with it. This subject-centred view of the physical world also affects the definition of corresponding spatial frames of reference. The subjective perspective begins with the idea that through the body the agent takes up a concrete position in the physical world.

> In every field of perception – science may generalize upon it and formalize it later, in whatever way – there is a kind of coordinate system, where, as the point of reference the subject's own body is the 'zero' point. In the orientation of human beings to the outside world there are rules which are related to the subject's own physicality.
>
> (Peursen 1969: 104)

The agent operates in the physical world 'by means of his animate organism' (Schutz and Luckmann 1974: 3). This world is invariably the point of departure for an agent's orientations and actions in the physical context. Thus the subjective, spatial frame of reference for the physical world is defined as a subject-centred one. It makes the orientation and location of material phenomena possible. This body-centred system of spatial coordinates comprises the following categories:

- the dimensions of 'left', 'right', 'in front', 'behind', 'up' and 'down' from the agent's line of vision;
- a point of origin where the above dimensions cross, formed from the perspective of the agent's bodily position.[25]

This means that the subject's 'here' at the time of the action is not an absolute fixed point, but a variable reference point of orientation in the physical world. In contrast to the spatial frame of reference in the objective perspective, the categories just mentioned are not homogeneous. Owing to the physical–biological conditions of the human body, they are not of equal value for the agent. 'Up' and 'down' are defined by gravity and do not change with human movement in the physical world. 'In front'/'behind' and 'left'/'right', on the other hand, change with every turn of the body. Hence they are more problematic as intersubjective categories of orientation. This difficulty is all the more significant in that it is precisely the horizontal categories that are all-important in relation to most practical action: they comprise the two-dimensional earth–space which is the basis of human activity.

Strictly speaking, every agent thus has a different point of origin for orientation. This potential problem is solved in everyday life by the fact that agents in the natural attitude adopt an 'idealization of the interchangeability of standpoints' (Schutz and Luckmann 1974: 60). That is, subjects assume that the other's point of origin can be adopted by themselves, and that the spatial view of the physical world can then be structured by the self along the same lines as that of the other. Yet while this solves problems in a face-to-face situation, difficulties are soon encountered in the spatial orientation to the physical world, if direct contact is no longer present.[26]

In everyday life, however, long-term points of origin are chosen subjectively. These are usually points which are particularly important to the agents in question, for instance their own houses. All other places in the context of action in the everyday physical world are located in relation to this base as a starting point. According to Bollnow (1980: 58f.), this becomes particularly apparent when people move house:

> When we move house, the world is re-formed in a new way in relation to our new home base. The same is true when we move to a new town or area. Relations with other towns are now formed in a completely new way: what used to be on the periphery moves into the centre, and vice versa.

We can thus assume that agents, in their subjective orientations to the physical world, refer to a series of spatial indications. Starting from categories pertaining to the body in relation to their current physical position, they focus on a subjectively defined point of

origin. We know from ethnographic and cultural-geographic literature that there are societies with regionally restricted pre-
industrial cultures, which have only *one* generalized subjective
spatial concept. The point of origin is frequently the chief's dwell-
ing or some important place in their spiritual beliefs or religion.[27]

A spatial frame for the social world

The social world is constituted in the life-forms of 'the Thou-
oriented I' and 'the speaking I'. Communication takes place
between agents in the Thou-oriented attitude, in terms of
expressed symbols (bodily movements, language, artifacts). It is
through such communication that agreement is tested regarding
subjective spaces, i.e. the way in which impressions are located in
the various provinces of meaning. Unquestioned 'coordinations'
of the subjective spatial frame of reference acquire intersubjective
validity. Those aspects of the subjective world which are, in the
reciprocity of perspective, mutually confirmed by interaction partners, should be seen provisionally as the subjects' valid categories
for orientation and ordering in the socio-cultural world. For
Schutz, adequate agreement is achieved only if the different
partners are using the same relevance system in an interaction
situation.

In sum: the constitution of the subjective world forms the basis
for relations with the other worlds. The basic categories for the
other spatial concepts must be derived from the nature of the
subjective world. The body of the knowing subject is a component
of the physical world and occupies a position in it, distinguishing
the particular nature of experiences from this standpoint. The
spatial dimension, therefore, also plays a decisive part as a
differentiation variable for the subjective world. And since the
social world is manifested through the partners during interaction,
it is on the one hand differentiated by the subjective and spatial
dimensions; on the other, the social world also structures the
subjective world and the material world of agents. For movements
in the spatial context are intentional, and their intentionality is
constituted against the background of a socially formed and
largely associational stock of knowledge.

A SEPARATE SPATIAL FRAME OF REFERENCE FOR SOCIAL GEOGRAPHY?

Our examination of the spatial problem has so far followed the ontologically different world outlined in the epistemologies of Popper and Schutz. We shall now bring the concrete actions of human beings into this investigation, and must therefore focus more closely on one area: the world of material artifacts, i.e. the materialized results of human actions. These cannot be assigned solely to either the physical–material world or to the immaterial social and subjective worlds. To assign them to the physical world would be inappropriate, because artifacts also preserve the meanings given to the acts which produced them. Artifacts are 'carriers' of subjective socio-cultural meanings. On the one hand they acquire a unique autonomy on completion of the act of production; on the other, they nevertheless point to the meaning of the act of production, and so preserve it. To assign artifacts solely to the social world would be inappropriate because they are material in nature, and therefore have a different ontological status from that of pure meanings and ideas.

In his action-oriented analyses of artifacts, Popper (1979: 112–13) appears to assign all products of human activity to world 3. If, however, we take his differentiation of the three worlds (the 'ontological differences of phenomena') seriously, material artifacts cannot be assigned unequivocally to world 3. Because of their very materiality, they cannot be located in the social world alone. In Schutz's action theory analysis (1972: 119ff.; 1974: 73), he describes artifacts as the most anonymous area of the *social world*, as 'like witnesses [they] refer back to the subjective meaning-contexts of an unknown manufacturer, consumer, spectator' (1974: 73). For Schutz, artifacts represent the subjective and social world as much as the actions of contemporaries. As I said, the difference is the level of anonymity involved: the inclusion of an artifact in a particular action is an 'anonymous social interaction' with its producer. I shall analyse this in more detail.

Just as the agent's body provides the material basis for an action, so the material aspect of an artifact is the precondition of the artifact's social meaning-content. But while, so far, the equation of the two is acceptable, an important difference is thereby obscured. The agent's body possesses a relatively autonomous consciousness, in which the intentions underlying movements are constituted.

Material artifacts, on the other hand, have long-term meanings which can only be interpreted within the framework of material compatibility. In addition, they are only available within the framework of certain agents' property rights. These are the reasons for the relative permanence of the meanings of material artifacts. The relative permanence of material artifacts is extremely important for human actions which relate to them, and for the location of the latter. The category of immobile material artifacts is particularly important in social geography, as artifacts give a quasi-permanent structure to the physical–material world, and its spatial dimensions, from the point of view of the social aspect. They form the fixed points of social interactions, to which agents must go if they have made them into the goal or means of their activities. The workplace is one instance. Patterns of social interaction in concrete social processes are spatially structured in relation to immobile artifacts' social meanings. This may be social interaction while using artifacts, or the use of artifacts in order to establish direct contact with interaction partners.

As we shall see in more detail later on, the spatial ordering of artifacts has diverse consequences for the social world. But if artifacts are analysed solely with regard to their materiality and their position in the physical world, and if their positions are not determined on the basis of adequate frames of reference in the subjective and social worlds, major difficulties arise for research in action-oriented social geography. In the 'objective perspective' of the 'spatial' approach, it was usually only the spatial patterns of immobile material artifacts (transport provisions, buildings, etc.) and the spatial distribution involved which were analysed. Similarly, numerous research projects on mobility sought or seek to analyse the processes of bodily changes of position in physical space. These research projects only use the spatial reference pattern of the physical world for locating and understanding movements. This means that only geometrical or (geo-)mechanical research is carried out. Just as little attention is paid to the social meanings of artifacts as to the intentions of agents moving in the physical world.

By contrast, genuine social research in the *objective perspective* should locate the meanings of artifacts under investigation in the social world. If this aspect is neglected, a central component of human life is perforce excluded from social research in human geography. It should accordingly be possible to grasp the complexities of the manifestation of the social aspect through the material component, order it appropriately, and take account of it in geographical explanations of social situations. In other words: situational analyses in human geography should take account of the social aspect, the physical–material aspect, and society as conveyed *via* the material aspect.

In the *subjective perspective* it is the courses of action which become the central concern of analysis, not the consequences of actions as such. Accordingly, the main interest of research is in the goals and intentions of actions, and in the social conditions for their provenance and realization. The goals may relate to either the physical–material sphere or the social sphere. Action orientations relating to these spheres should first be understood from the agent's perspective, and located by means of adequate spatial concepts. But even if the agent is striving for goals in the social world, he or she often has to use means which have their own symbolic meaning, even though they are also located in the physical context.

The question now is whether a separate spatial frame of reference for the 'artifact-world' is necessary for understanding the social and material aspect of artifacts, as a preliminary condition for the explanation of problematic social processes, or whether a comparison of frames of reference for the social and physical worlds is sufficient. Our deliberations so far seem to favour the construction of a separate reference pattern, since artifacts cannot be assigned solely to the physical or to the social world. Yet this solution is fraught with difficulty. It involves the risk of reducing the social to the physical, or of ignoring the physical altogether. Because of the ontological difference in the phenomena to be located, the two areas cannot be classified according to like criteria. For this reason, the social and physical dimensions cannot share a single point of origin.[28] This means there is no point in trying to find *one* spatial reference pattern for both worlds.

Ideally, what we want are criteria or rules which make it possible to compare the locations of artifacts in the physical world (their material component) and in the social world (their persistent meaning) in all their complexity. But we cannot develop these detailed criteria or rules here. Nevertheless, their development seems to be the most hopeful way of avoiding spatial reductionism in social and cultural geography. The objective perspective should thus involve a constant comparison of location patterns in relation to the frames of reference of the physical and the social worlds. The subjective perspective should take into account the position of the meaning in relation to the frame of reference of the subjective world.

If *social* geography is to live up to its name, it must be able to analyse, understand and explain appropriately those elements of the social, physical and artifact worlds which become relevant in courses of action. In doing so, it must take account of the reciprocal relationships between them and of their respective meanings for different societies. It is only in this way that solutions proposed by social geographers for problem situations could have success or meaning.

In the following chapter I shall be examining in some detail the ways in which sociologists, on the periphery or outside the field of action theory, integrate the spatial arrangement of material phenomena and facts. More importantly, in this context, I shall attempt to clarify the more general sociological preconditions for the development of an action-oriented social geography. As I counterposed an action model to a traditional spatial model at the outset, arguing that the reductionism of the latter had to be replaced by different spatial concepts which were consonant with an emphasis on action, these spatial frames of reference will also be borne in mind.

Chapter 7

The space of social theory

Before we examine the importance attached by classical action-theory sociologists to the spatial arrangement of material objects and artifacts, we shall first search for points of contact between non-action-oriented sociology, and the development of an action-oriented social geography.

THE VIEWPOINT OF SIMMEL

Following Simmel (1903), we can distinguish three basic aspects of the spatial dimensions of the physical world which are important for an action-theory analysis.

- the exclusivity of the traditional geographical space;
- the mobility/immobility of objects and artifacts in this space;
- agents' nearness or distance from each other and immobile objects.

Simmel's '*exclusivity of space*' means that if an object is considered only from the point of view of its location on the earth's surface, and all its other characteristic dimensions are ignored, it is always unique: at any given time only one object can occupy a particular position. 'It is because every [object] occupies a different part of space that there are *several*, even though they are absolutely identical' (Simmel 1903: 29). Uniqueness of position at a given time is also a characteristic of the agent's body. The bodies of two agents cannot occupy exactly the same position in the physical world. Even if the agents' bodies were in other respects absolutely identical, they would be distinguishable by their position in the context of the physical world.

Apart from the unique position they occupy (which could, in theory, be a fixed one), agents and objects can also be distinguished by their physical '*mobility or immobility*'. We start out from the basic premise that the agent can change position. Further, we can make a distinction between objects that are immobile in relation to the actions and those that are, at least potentially, mobile. Simmel (1903: 39) comments on the social importance of this distinction: 'Whether a group, or certain separate elements of it, or important objects of interest to it are absolutely fixed or [. . .] indeterminable clearly influences its structure.' In the context of action, this idea can be taken a step further: whether the material objects which become important for certain actions are mobile or immobile influences the structure of the empirical course of an action from the standpoint of the physical world.

Simmel points out that the physical immobility of an object relevant, say, to the means for achieving a goal, 'produces certain forms of relationship which are grouped around that object' (1903: 40). The immobility of a material object relevant to action thus obliges agents *to go to it* if they wish to utilize it in achieving a goal. This means that certain social relationships must be ordered around immobile material objects. The spatially defined location of the immobile material object thus becomes a socially important pivot of human interactions. Such a pivot may be represented by a single building or a whole town, according to the observational scale used. At all events it is clear that this fixed spot in the physical world becomes 'a pivot for the relationship and the social context' (Simmel 1903: 41) for all agents integrating this immobile material object into their actions.

Simmel argues that the 'pivots' may also acquire symbolic content. They may, for instance, exert a certain force, as in religion, which 'reawakens an awareness of belonging in agents whose religious needs have long been buried through isolation' (1903: 41). We shall later consider in more detail this aspect of the symbolic content[1] of immobile material objects and their importance for the agent's orientation in the social world and thus also in its spatial context. For the time being it is enough to note how immobile objects relevant to actions can structure social relationships.

I turn now to Simmel's third category: 'the physical nearness or distance of persons who relate to each other in some way' (1903:

46). This clearly derives from the 'axiom' of the exclusivity of positions in the physical world. Since objects and the bodies of agents cannot occupy the same spatial position at a given time, there is always a distance between them which can be characterized as nearness or distance in relation to a uniform spatial measurement. Simmel concentrates on the distance between agents, especially the way in which a social relationship between agents changes according to the geographical distance between them. His analysis could possibly be adapted to the relationship of agents to immobile objects and artifacts relevant to action (although Simmel does not do so himself). Here, however, I shall limit my focus to changes in the nature of social relationships arising from changes in the spatially defined distance between the bodily positions of interaction partners.

These 'will change in character according to whether the participants are "spatially" contiguous or separated from each other' (Simmel 1903: 46), in the sense that the necessity of overcoming distance becomes a factor in 'the differentiation of relationships'. Only two forms of social relationship are unmodified by changes in physical distance: 'purely objective-impersonal relationships, and those involving overwhelming intensity of feeling' (1903: 47). He gives 'economic or scientific transactions' as examples of the first type, since their content can be expressed solely in logical and written forms, thus obviating the necessity for bodily nearness in the physical world. As examples of the second type he gives religious feelings and love relationships, where physical distance can be overcome by means of imagination and emotional devotion. The more a relationship diverges from these two extreme types, the more the nature of social relationships can be changed through the factor of geographical distance and the more the interaction partners need spatially defined nearness for the preservation of their social relationship. Finally, Simmel suggests that all levels of relationship could be arranged on a scale between the two above-mentioned extremes. The criterion for such an arrangement would be 'the measure of "spatial" nearness or "spatial" distance which either furthers or tolerates the association of given forms and contents' (1903: 47). Simmel undertook no empirical tests of these hypotheses. But he did suggest two preliminary ideas which could be useful for empirical research in an action-oriented human geography.

First, the spatially defined distance between bodily positions which can support a social relationship 'depends on the measure of capability of abstraction' (Simmel 1903: 47). For the less capability there is, the less able the agent is to relate *socially* to a person who is *physically* distant. If this capacity for abstraction is absent, social relationships involving only short physical distances take on a form such that everybody in close proximity becomes the focus of a complex and emotional attitude: one that is either very friendly or very hostile. The agent who lacks this capacity for abstraction perceives his pattern of social relationships as the direct consequence of physical distance.

Second, most forms of 'association' based on the physical proximity of bodily positions are more intense than those that are maintained at substantial physical distances. This is because 'idealizations' are more likely to be correct, when their reference 'points' are close at hand. Physical proximity enables us to test the ideas we have of others to a far greater extent than when we are at a physical distance. Social relationships in close physical proximity are therefore also subject to more searching scrutiny by interaction partners. For this reason, Simmel (1903: 55ff.) thinks that *direct democracy*, which is characterized by constant supervision of central government by the electorate, can only work if the size of the territory involved is strictly limited.

THE SOCIAL IMPORTANCE OF MATERIAL ARTIFACTS: DURKHEIM'S SOCIOLOGY

Durkheim's ideas on the importance of the spatial dimension of the physical world for social reality are elaborated in *Les formes élémentaires de la vie religieuse* (Durkheim 1985). Basic categories of thought such as 'space', 'time', etc., the capacity for, and nature of, classification, and the formation of categories, or 'spatial thinking', are dependent on the specific structure of a society. The particular form of the thought category, 'space' is geared towards 'intersubjective understanding with regard to spatial relations, and the spatial symbolization of socially important things and events' (Konau 1977: 20). For Durkheim, this always depends on the nature of the social world from which it emerged: 'these concepts express the way society represents things' (Durkheim 1985: 626).[2] On the one hand, the definition of the spatial frame of reference for location and 'intersubjective understanding regarding spatial relations' depends on the particular world view of a society. If this is determined primarily by religion, the dimensions of the physical, spatial frame of reference will be influenced by the relevant mythology or dogmas.[3] On the other hand, the spatial ordering of the immobile material substratum, the world of immobile material artifacts in a society, is also dependent on the social structures of that society. Durkheim sees the material substratum of a society as a central social fact (*'fait social'*). He defines this more precisely as 'any way of acting, whether fixed or not, capable of exerting over the individual an external constraint [and] [. . .] having an existence of its own, independent of its individual manifestations' (Durkheim 1982: 59).

In addition to the material substratum of immobile material artifacts, he draws attention to immaterial, institutionalized patterns of action.[4] He thinks we must analyse immaterial patterns of action in the same way as we analyse material things. This is a mistake, for reasons I shall discuss in more detail below. *Patterns of action should not be analysed as material artifacts, but material artifacts with their social content should be analysed as patterns of action.* If this is done, other ideas Durkheim has about the material substratum can be integrated into an action-oriented social geography.

For Durkheim (1982, 136), things which form the material substratum of a society are 'the products of previous social activity'. This applies both to their different content (buildings, roads, tools, etc.) and to the nature of their distribution in the physical world. Although material things themselves do not 'release any vital force', they do exert 'to same extent [. . .] an influence upon social evolution whose rapidity and direction vary according to their nature' (1982: 136). Thus the nature of the material substratum and its arrangement in geographical space becomes just as important for social structure as the immaterial action patterns of a society. Although 'in fact the specifically human [sphere][5] remains as the only active factor' (1982: 136), material artifacts as 'crystallized forms of social reality' (1982: 82) also exert a constraint upon actions, which is 'far from their being a product of our [the user's] will' (1982: 70). I shall now analyse more closely this curious dialectic between material artifacts and individual action.

The socially relevant aspect of artifacts is evident in the fact that they 'are like moulds into which we are forced to cast our actions' (1982: 70). In other words: material artifacts can direct actions.

Every material artifact presents a certain means–end relation to agents, which they are not free to interpret as they like. Indeed, for the most part their interpretations are very restricted. They have to accommodate the intentions of the inventor and builder of the artifact. Material artifacts thus acquire the importance of *instrumental institutions*, which also direct actions, and become relevant to social structure in a way that goes beyond mere usefulness.

So the social content of material artifacts is important in two ways. First, a social meaning is preserved within them. The user's action must adapt to this and be more or less oriented towards the intentions of the producer of the artifact, if the action involving the artifact is to be successfully executed. Second, it becomes clear that the user indirectly enters upon an anonymous social interaction with the producer. Both these aspects are elements of the social world, and should be structured and located on the basis of a social, spatial frame of reference. Meaning should be established according to the artifact's particular position in the socio-spatial frame of reference, and interaction seen as the relation between positions of meaning. The material aspect should be located in relation to the categories of the physical–spatial frame of reference.

As I have indicated, reinterpreting Durkheim's approach is important for the construction of action theories in social geography. Read together with the epistemological differentiation of 'space', and the ideas of Simmel, Durkheim opens up a new dimension for geographical analysis of forms of association. With Simmel, we saw that immobile artifacts structure actions (according to their particular goals and the arrangement pattern of immobile objects) within the context of the physical world. Association takes place *around* those immobile artifacts which become means for action. Durkheim adds a further insight here when he points out that social meanings are also preserved *in* material artifacts as patterns of action, and therefore also structure action spaces in the physical world from the social aspect. Every empirically ascertainable action which integrates an immobile material artifact into its performance is thus determined by a simultaneous double orientation. First, towards the physical world *via* physical–spatial categories, and second towards the social world *via* social–spatial meaning-dimensions.

The reference area of immobile material artifacts is important for the explanation of the differentiation of the socio-cultural world in its spatial context, for the following reasons. Such artifacts give relative permanence to social meaning, possibly lending stability to the social world. In addition, their immobility gives relative permanence to arrangements in the physical world, possibly increasing the durability of action spaces in that world. The result of both factors is that certain courses of action become so embedded in the physical and social worlds that their long-term continuance is synonymous with the survival of a particular social system.[6] In an otherwise rapidly changing social world, it is possible that artifacts play as important a role as immaterial traditions in providing relative security in action orientation through routinization.

Again, given my emphasis on subjective agency and argument for the necessity of an action-oriented social geography, it is particularly necessary to stress the importance of immobile material artifacts in the reproduction and change of existing social conditions. As I said in Chapter 1, these establish constraints on the nature of subjective agency. However, we have shown in this chapter that these immobile material artifacts are not, of themselves, reducible to 'space' as a cause.

SYMBOLIC CONTENT OF PLACES

In developing an action-oriented social geography, there is another topic of sociological interest we need to consider. It is the *social* meaning of the physical world and the world of material artifacts, and (related) the symbolic significance of places and place-names. This topic has recently been revived in urban sociology.[7] It turns our attention to something we have not yet touched on: namely, how natural objects are invested with symbolic meaning. Agents may assign a meaning to an object or point, even though these have no social meaning in themselves (stones, trees, mountains, etc.). Similarly, socially constructed artifacts may be invested with a symbolic meaning which need not coincide with the rationale underlying, and preserved in, their construction.

The question is: What significance does this symbolic investment have, and does it bear on the significance of spatial differentiation in social worlds for concrete courses of action?

By his analysis of the 'rendez-vous', which 'denotes both the meeting itself and the meeting-place' (1903: 43), Simmel provides an important pointer to the symbolic content of places and place-names, and to its specific significance for action. His analysis in fact gives us the basic structure of the processes through which certain places, artifacts at a particular point in space, and place-names, acquire symbolic content. The immobility of material artifacts structures the action spaces in the physical world in a unique way. They become spatially fixed pivots for social relationships. Through habitual action, this or that place 'acquires a special air of security for the consciousness. It usually has stronger associative powers for the memory than time, because it is more vivid to the senses' (1903: 43). In other words: a special meaning is transferred to the place, the artifact or the place-name where or through which the action took place, 'so that in the memory the place tends to be inextricably linked with that (action)' (1903: 43). Simmel uses here the examples of 'church', 'town' as well as 'rendezvous', all of which he sees as pivots of social relationships which can be appropriately located in physical space. In all three cases the value accorded them gives agents interacting at those places 'an awareness of belonging' (1903: 41). 'The place remains the pivot' (1903: 43), and all participating agents consequently invest it with symbolic value, 'around which individuals' remembrance wraps interrelations which have now become abstract' (1903: 43). For Simmel, the symbolic plane of social communication is of the utmost importance for the development of the 'awareness of belonging' (Simmel 1903: 41) to a social organism, or more precisely, for territorially limited social integration.

The transference of the sensory traces of actions to the place where they were performed is expressed in the name given to them. A house with a proper name gives those living in it a feeling of individuality, of belonging to a *qualitatively* fixed point in physical space (Simmel 1903: 43). Generalizing from this we can say that an idea is associated with the name of a particular place or geographical unit (settlement, region), which expresses the individuality of the way of life of the people living there. For Simmel this is important in relation to the social world, since the individuality a place has for particular persons or groups can either further or hinder social relations extending beyond that particular place. Whether such relations are furthered or not depends on the nature of the symbolism inherent in the name. This symbolism can have either a general or a local form. The Vatican means something for the Catholic church in general. 'Love of one's native home' is local. Obviously the first form favours social relations which extend beyond one's local place of habitation, while the second is more likely to inhibit them.

Memory features in a different way in the work of Maurice Halbwachs. Following Durkheim, Halbwachs analyses the relationship between the individual and the collectivity. His basic theory is that the memory and judgements of agents are always subject to specific social conditions

> which could be said to provide the framework in the context of which agents think and remember: within collective notions. Such notions vary, however, from society to society, from class to class and from group to group, and are characteristic for a particular society, class or group.
>
> (Maus 1967: 7)

In Halbwachs' view, individuals orientate themselves in their actions towards the collective memory, and are therefore determined by it. More importantly for my purposes, he analyses the 'collective memory' in terms of its temporal and spatial dimensions and differentiations. In the following I shall concentrate on how the spatial aspect of the collective memory influences action.

The collective memory is the intersubjectively available knowledge of events and objects within the framework of a particular group. 'Each (individual) memory is a "view point" on the collective memory, that this view point changes as my position changes, that this position itself changes as my relationships to other milieus change' (Halbwachs 1980: 48). Spatially, the material world, with the symbolic memories it embodies, is the sphere in which the individual agent, from his or her position in the social and physical worlds, enters relations.

The material milieu of a particular geographical unit in which the collective memory is preserved is seen by Halbwachs (1980: 129) as a 'part of society [. . .] which recalls a way of life common to many men'. Even though the elements of the material milieu do not possess the faculties of agents, 'we nevertheless understand them, because they have a meaning easily interpreted' (1980: 129) because they have been formed by a group of which we are a part: 'place and group have each received the imprint of the other' (1980: 130). In this reciprocal relationship, collective memories are transferred to the physical surroundings of the action. The continuum of the physical and social worlds is important here in that it determines the uniqueness of every constellation of material artifacts. By the same token, each agent occupies a unique position, in terms of his or her projected actions in space, in relation to his or her constellation.

As the social world changes, the symbolic meaning of its material milieu changes also. And as the material milieu changes so, according to Halbwachs, does the group. The significance of the symbolic content of immobile material artifacts for security in social orientation thus becomes clear. Halbwachs (1980: 128) puts forward the theory 'that mental equilibrium was, first and foremost, due to the fact that the physical objects of our daily contact change a little or not at all, providing us with an image of permanence and stability'. The intentions of our predecessors, taking concrete form in a material spatial arrangement of artifacts, are also passed on through this material milieu, and 'the power of *local* tradition derives from the thing whose image it originally was' (Halbwachs 1980: 137). It is thus through the world of artifacts that the socio-historical context of actions is differentiated within the physical world. We may therefore say that Halbwachs sees the symbolic contents of the material substratum of a society and its spatial immobility as, first, providing continuity in the face of social change. Second, for Halbwachs they pass on collective memory from one generation to another.

Halbwachs' hypotheses have typically been invoked by sociologists rather than empirically tested. For an action-oriented social geography the ideas discussed above, if they are empirically valid, could become important in the explanation of the material mediation of a spatial frame of reference in the social world. His hypotheses could be of particular importance for research into urban life, allowing a more detailed description of the social character of the material environment. This, in part, Treinen attempted.

In a summary of the main theories of Simmel and Halbwachs, Treinen starts from the idea that symbolic 'relatedness to a place is not a special phenomenon and does not concern the way man comes to terms with his physical environment. It is rather the specific form of a more general problem: the symbolization of human relationships' (1974: 237). Agents associate their personal experiences with physical locality and the objects and artifacts located there. From this, Treinen (1974: 238) concludes that a symbolic relationship with a place is formed 'when the place is the linking element in the social actions of the members of a local social structure', and that relationship to place is manifested 'after a detour *via* identification with a place-name, which represents the symbol for this category [of place].

Accordingly, Treinen focused his investigations on the empirical recording of various meanings of place-names, and the reconstruction of the process which led to this particular form of symbolization. He discovered that the following factors are decisive in this process. If social actions are frequently repeated in the same physical situation or in the same situational context, the elements of either may become so closely associated with the action that one of them becomes the symbol of the action. This is often the (place-)name of the physical–material aspect of the action situation. The specific form of the meaning of the symbol depends on the intensity and frequency of the action. 'If the nature of the relations between the participants changes, it is possible that the symbols concerned remain but with a changed meaning' (Treinen 1974: 239). The specific meaning which a particular symbol has for agents, and the ideas and emotions they associate with it, depend on how they interact in the situation which the symbol represents. Moreover, whatever the form of symbol, according to Treinen's theory, it is an element in the 'frame of reference for action orientation'.

So the symbol represents the characteristics of the social relations between the agent and the other participants in the same action situation. The meaning of the symbol accordingly becomes the expression of the way in which an agent's identity is closely linked to situations. He or she uses the symbol and its meaning for orientation, for those acts which relate to the symbolically represented situation. This of course can be tested or empirically investigated. Treinen has brought Simmel and Halbwachs down to earth (as it were).

It remains to see how social theories of action bear on geographical space in more general terms.

Chapter 8

The space of social action

In my examination of the work of Pareto and Weber thus far, I have concentrated on their models of rational action, within the framework of their general theories of action. I shall now turn my focus to the level of importance they attach, in explaining forms of associations, to spatial aspects of the physical world and to the material mediation of artifacts. I am seeking pointers which will be of use in the development of an action-oriented theory of social geography. Pareto and Weber too, discussed spatial patterns in the arrangement of material objects and artifacts, together with their symbolic encoding. They did so in relation to the performance of action. Initially, I examined Pareto and Weber together with rational choice theory. This discussion will follow suit in that I shall also, in this chapter, return to economics, via the spatial dimension in rational choice theory.

PARETO: THE PHYSICAL WORLD AND RELATIONS TO PLACES

In discussing logical actions, Pareto (1980) implicitly stresses the importance of spatial differentiation in the physical aspects of action situations. In his theory, a logical action accords with the objective conditions of the situation, and I have shown that action situations will of necessity vary from the spatial point of view. Pareto, however, is not clear about the relation between actions and objective conditions. It may mean that the agent acts upon objectively correct knowledge. Or it may mean that the physical world always plays a role if a material artifact is used as an end or means; and if the physical world is objective, so is its spatial dimension.

The agent is always confronted with the question: Do I go to it, or can it come to me?

For Pareto non-logical actions constitute most of social reality (Pareto 1980: 20ff.). The significance of spatially definable elements is touched on in connection with emotional relationships to places. Pareto refers to the determinants of non-logical actions as 'residues'.[1] These do not result from correct observation and reasoning, but stem from the emotional sphere.

He terms one such class of residues 'persistence of people's relationships to places'. He divides this class into three: relationships to family and community; relationships to places; and relationships between classes. Pareto (1980: 114ff.) derives the importance of emotional 'relationships to places' for action orientation from 'relationships to family and community' especially.[2] Thus we shall concentrate on these two divisions.

According to Pareto, the emotional relationship to the family develops from children's dependence on their parents over a long period of time. This leads to the creation of powerful residues. It is then no longer rational considerations which determine actions, but emotions evoked by the personified family name. Similarly, in the case of actions relating to village communities: the emotional relationship to the name of the village determines the orientation of actions connected with this unit. The relationship to places as a residual category of action orientation is thus nothing other than the result of the transference of the emotional content of social relations on to the name of the physical spot where the corresponding relations took or take place. Thus the emotional relationship to place-names can be seen as confusion between place-names and the emotional content of relationships to the family and the community: 'relations to places combined with relations of family and groups [. . .] to form a unit sum of residues' (Pareto 1980: 131). For Pareto the term 'motherland' denotes simply an administrative and territorial unit within which one was born and spent one's childhood. For agents it becomes, like other such expressions and place-names, an emotionally charged symbol for the agent's relationships to family and community which exist in that place. Thus place-names influence orientation for all non-logical actions relating to contexts such as the defence of one's homeland, duties towards the local community in which one believes, etc. War is therefore nothing else but a sum of non-logical actions.

Similarly, patterns in the spatial arrangement of material objects and artifacts are invested with symbolism, qualitatively differentiated. We can thus put forward the hypothesis that the physical world is also symbolically structured by the social world, and that this symbolically 'charged' physical world also structures social actions.

WEBER: THE PHYSICAL WORLD AND MATERIAL ARTIFACTS

In order to draw out Weber's assumptions about the physical world and the social significance of material artifacts in his action theory, we need a more comprehensive framework of analysis than that provided by his own typology of action. This framework can be derived from the three-world model. But to do this, we need to situate the three worlds in relation to Weber's action theory.

Let us begin by referring to Weber's definition of social science, and of sociology in particular. He describes the field of research of the social sciences (1951: 429) as 'not any kind of "inner state" or outward behaviour, but: "action"', i.e. those human activities which 'relate in a subjectively rational way to the external world, in particular to the actions of others' (1951: 429). It is thus already clear that mental states 'as such' ('inner state') are disregarded here in the same way as 'irrational' activities are disregarded. For Weber, the only subjective aspects that are important are those which relate to the social world (actions of others) and the physical world (external world in general).

For Weber the relation to the material external world is not a subject matter of interpretive sociology, but certainly for other social sciences: the objects of the material world have by 'both the actor and the sociologist [to be accepted] as data to be taken into account' (Weber 1968: 7). The 'external world which is "meaningless in itself", things and processes in nature' (1951: 431), acquire relevance for the social sciences in 'their role as "conditions" and "consequences" towards which purposive-rational action is oriented' (1951: 431). To put it another way: all facts belonging to the physical world are in 'all sciences of human action [. . .] considered as [occasion],[3] results, favouring or hindering circumstances' (1968: 7). They should be noted if and only if they 'relate to action in their role either of means or of end; a relation of which the actor or actors can be said to have been aware and to which their actions have been oriented' (1968: 7). Weber (1951: 430) sees the task of the social sciences as 'explaining clearly' which kind of rational action intervenes successfully or unsuccessfully in the in-itself meaningless physical world, and which consequences result from that action for the further actions of others: by occasioning them, favouring them, or constraining them.

Weber gives a complex analysis of the relationship of human action to the world of material artifacts. He distinguishes clearly between inanimate objects which are not meaningless, and those which have 'no intended meaning'. He emphasizes that an artifact 'can be understood only in terms of the meaning which its production and use have had or were intended to have; a meaning which may derive from a relation to exceedingly various purposes. Without reference to this meaning such an object remains wholly unintelligible' (Weber 1968: 7). This meaning becomes a component in the agent's orientation: 'He is interested in the expectations he has of the conduct of these artifacts, which are of practical importance to him' (Weber 1951: 457) – a similar attitude to that which the agent has towards social institutions.

Within the framework of his analysis of the Western rationalization process and the division of labour, Weber (1951: 473) points out that the agent's world consists more and more of artifacts, which, although they can be rationally understood and controlled, are less and less often produced by the user. This means that as the specialization of the individual's activities increases, more and more areas of his or her action are defined by constructions which were produced by others for a particular purpose.

Although Weber does not explicitly express this idea, his argument presupposes that the world of material artifacts is of paramount importance for the various forms of association and the division of labour. The more consistently someone is 'exposed' to an environment of human artifacts, e.g. in the various forms of urban life, the more his or her activities are specialized. The more agents relate in their activities to artifacts produced by others, the more they restrict 'the importance of charisma and of individually differentiated action' (Weber 1968: 1156). Moreover, for Weber, specialization and the reification of social relations are two aspects of the rational standardization of the 'obedience of a plurality of men' (1968: 1149), both of which limit the influence of charismatic authority. 'Rational conditioning and training celebrates its greatest triumph' (1980: 686) when the actions of human beings are totally adapted to the requirements of material artifacts, so that courses of action acquire 'a new rhythm through the functional specialization' (1968: 1156). Although Weber is referring here primarily to the 'modern capitalist factory' (1968: 1156), one can generalize from this example to most forms of human adaptation to the proliferation of (immobile) material artifacts, including those encapsulated in urban areas.

We can now distinguish in Weber's theories three categories of action, which bear on the three (physical, subjective, social) worlds respectively.

1 *Social action*: intelligible action which subjectively links objective social meaning subjectively with an activity, and is oriented towards a meaning of the actions of others. Social action divides into many different forms of institutionalized action and represents a specific area of research in sociology: relations within the social world.

2 *Action relating to artifacts*: intelligible action rationally oriented towards artifacts, which ties artifacts, as means or goals, into action. For Weber artifacts have a social content, expressed through the anonymous social relationships they create, which in turn bear on actions.

3 *Action relating to the physical world*: intelligible action which subjectively links objective social meaning with an activity, and which involves the physical world as a condition of action. However, the physical world is seen merely as occasioning, favouring or constraining agents in relation to the intended activity. Although agents orient themselves towards the physical facts, they are not able to grasp their actual meaning. The meaning reference remains one-sided: the relationship is between the agent's subjective attitude, which has meaning, and the physical world which has no meaning in itself.

The first topic is the main one for sociological research. Weber assigns the remaining two to other social sciences, though without specifying precisely which one he means. They are assigned to the periphery of sociological research proper. Topic 3 provides the 'data to be taken into account' and should therefore be seen as a kind of initial phase. The status of topic 2 is not clarified. An action-oriented social geography could certainly take over topics 2 and 3 as its field of research, however. An investigation of relations to the physical world could, under certain conditions in the social world, be research oriented towards human ecology. Relations to the world of artifacts could be seen as a field of investigation in the social world in all its physical diversity. Even from this outline it should be plain, however, that such investigations could never be undertaken without adequate knowledge of the social world and the theories pertaining to it.

As we have seen, Weber suggests three ways in which artifacts are important. First, they are important within the framework of the 'evolutionary' process of rationalization in the Western world. Second, they are a limiting factor for the sphere of influence of charismatic authority. And third, they circumscribe 'individually differentiated action'. The first aspect can be understood as a reformulation of the topic area of 'alienation' as encountered in the theories of Marx, and the two others as representing its consequences. The first consequence bears on what Durkheim and Halbwachs have identified as the stabilizing factor of artifacts in social change. As to the circumscription of 'individually differentiated action', no doubt for Marx the most important negative consequence of alienation, we could consider these points: the fact that producers and users of artifacts are rarely the same people is not always negative. Nor is the fact that individual action is circumscribed. For both of these can lead to certain freedoms and advantages. On the other hand, there comes a crucial point where the meanings preserved in artifacts are no longer compatible with agents' physiological toleration and subjective intentions. It is then very probable that the users will become both physically and psychologically ill, and unable to cope with the consequences of alienation. The 'inhospitableness of our cities', as Mitscherlich (1970) puts it, acquires thereby a new dimension. This problem area should be integrated into an empirical theory of social geography.

Other areas of research which this analysis illuminates are those of regions and ecology. If the topics discussed above are to be integrated into an action-oriented social geography to good effect, it should be understood that immobile material artifacts represent the most important elements in which social world meanings are persistently expressed in patterns of spatial arrangements. Since the way they are distributed across a spatial continuum varies enormously, their precise investigation could lead to an indication of the reasons why social systems in different regions remain distinct from each other over long periods of time. This is probably one of the most important determinants of regional differences.

Turning from regional difference to ecology: we start, of course, from the fact that both the socio-cultural world and the physical world are spatially complex in nature. It has to be established what kinds of actions can intervene successfully in the physical world under which social conditions. And finally the consequences of such interventions for further actions of others have to be analysed: are they occasions [*Anlässe*], furtherances or constraints?

SPATIAL PATTERNS AND RATIONAL CHOICE THEORY

Leaving Weber and Pareto, we turn now to the use of rational choice theory for an action-oriented social geography. In economics, rational choice theory is characteristically related to action situations where the means for achieving goals are scant. The decision principle is this: 'Exploit the given means for the greatest return, or: achieve a given goal with the least effort' (Gäfgen 1974: 102). When the spatial dimension of the physical world is taken into account in a calculation, rational choice-theory transmutes itself into the location theory of economic institutions, of immobile material artifacts of production.

The choice of location for individual businesses is always a problem of finance. The question has to be resolved of 'whether, from the point of view of efficiently satisfying the wants of a given group of persons, it is rational to locate a certain enterprise with a given productive [goal][4] at one or an alternative site' (Weber 1968: 103). The particular criteria which influence the decision depend on the nature of the given goal, of the intended product. In the case of agricultural production there is the added factor of the nature of the soil,[5] and in industrial production there is the question of the availability of suitable raw materials and the proportion of weight lost in a comparison between the raw material and the finished product, etc.[6] In the service industries the main consideration in choice of location is the frequency of demand for the service on offer[7] at a particular place.

Similarly, the classification of the spatial dimension is primarily concerned with the physical distance between the positions of means which are relevant in the context of economic action. The continuum of the physical world becomes important in relation to distances; these are not geographically measurable distances, however, but *functional distances*. That is to say, distances that are measurable in units of cost-effectiveness.[8] In this way geographical distance is transformed by socio-economics into an element of cost calculation. The optimum location for the pursuit of a given goal is seen as that where the fixed (transport) costs of procuring the means of production and retailing the finished product are lowest. In location decision theory, through the transformation of geographical distance into a cost scale or economically functional distance, distances between positions in the physical world become a cost factor towards which logical decisions about location are oriented. No other aspects may play a role in the logical action orientation in decisions regarding location. They would otherwise have to be identified as evaluative, affectual, traditional (Weber) or non-logical (Pareto).

SPATIAL PATTERNS AND NORM-ORIENTED ACTION

Keeping to the logic of retracing theories already outlined for their bearings on action and space, I return now to Parsons. Konau (1977: 184) points out that for Parsons the term 'space' has different meanings at various stages in the development of his theory of action. I noted that in *The Structure of Social Action* (1937) Parsons investigates the importance of the categories of the mechanical–Euclidean concept of space for the development of a formal model of action. In *The Social System* (1952) he analyses the importance of 'space', in the territorial sense, in the development of social existence. Finally in 'The principal structures of community' (1960) he analyses the relevance of the spatial continuum in shaping concrete societies. In the following analysis I shall be concentrating for the most part on reconstructing the social significance of spatial patterns of arrangement, and considering problems associated with this. Once more, I shall be using the ontologically differentiated three-world model and the spatial frame of reference appropriate to each world.

Mechanical–Euclidean space and action theory

In *The Structure of Social Action* Parsons sees the different spatial dimensions contained in the mechanical–Euclidean concept of space (in contrast to the temporal dimension) as irrelevant to action theory: 'While the phenomena of action are inherently temporal, [. . .] they are not in the same sense spatial. [. . .] That category is irrelevant to the theory of action, regarded as an analytical system' (Parsons 1937: 45). If we take that last qualifying phrase ('as an analytical system') as an indication of Parsons' predilections about a formal model of action, we can interpret the whole statement as follows. With every action there is a certain time-span which stretches from setting-up of the goal, *via* the choice of means to achieve the goal, to the realization of the action. On this abstract level, however, neither the mechanical–Euclidean spatial concept nor the spatial dimension of the physical world plays a role in the formal representation of the ideas corresponding to this time-span. An understanding of the formal structure of social actions is thus not dependent on the spatial representation of the physical world, as it is in mechanics. Were we to use mechanical categories in representing courses of action, we would only be able to describe the physical aspects of actions (arm and leg movements, etc.) and not their intentional aspects. Parsons' statement, at it simplest, means that the action sciences cannot be a subdiscipline of physics. As social science disciplines, they can only explain actions in relation to their meaning. Parsons is not implying, however, that the spatial dimension of the physical world plays no role in empirical courses of action, for 'the events of actions are always concretely events in space, "happenings to" physical bodies or involving them' (Parsons 1937: 45). The idea that the physical bodies to which actions relate, or which are integrated into those actions, occupy a particular position in the physical world in the same way that the body of the agent does, indicates that patterns of spatial arrangement which correspond to action are always an important condition in concrete courses of action.

Regarding the concrete action situation, Parsons accepts the distinction between the 'conditions' and 'means' of the action. 'Conditions' refer only to those elements of the situation 'over which the actor has no control, and "means" over which he has control' (Schutz and Parsons 1978: 12). If we disregard for the moment the specific social aspects of these different conditions that agents have at their disposal, then objects which might become means, but which at the time of the action are not within physical reach of the agent, are the 'conditions', and those within agents' reach are (potential) means. But if we include the social elements in our argument it immediately becomes clear that many of the objects within physical reach of agents are in actuality not necessarily within their sphere of control, even if they are only their neighbours' gardens.

The physical world and social interactions

If human actions relate to the physical world, including the agent's body, then according to Parsons (1952) the agent applies two different criteria of orientation: one classificatory, the other relational. 'By classificatory criteria is meant those which orient the actor to the object by virtue of the fact that it belongs to a universalistically defined class which *as a class* has special significance to ego' (1952: 89). That is to say, the agent divides all objects in the physical world into particular classes. He subsumes those objects under general classes, which already have specific significance for him or her. In this way a first structuring of the action situation is achieved. Parsons uses 'sex' as an example of this classificatory criterion, by which a particular person can be assigned to the class of 'female person' or 'male person'. 'By relational criteria [. . .] is meant those by which the object as particular object is placed in a specific significant *relation* to ego and thus to other significant objects' (Parsons 1952: 89). The second category of orientation thus helps agents to determine the significance of physical elements of the situation in relation to the purpose of their actions. In this way the agent establishes a relation between the importance of the goal of the action and of the physical objects involved in the situation. Just as 'sex' is a classificatory criterion, 'mother' is a category which establishes a relation of significance between myself and a person belonging to the class of female persons.

In contrast to the first criterion of orientation, in the case of the second the spatial dimension of the physical world becomes relevant, and always (as Parsons 1952: 90) puts it from the 'territorial standpoint' of the agent's body. He considers this territorial location in the spatial continuum of the physical world as an important aspect of orientation. For the agent always has to relate to other objects and their physical positions from a given bodily position. If we now also introduce the agent's position in the social world, particularly in the economic and legal subsystems, into his or her meaningful relation with material objects, the differentiation between 'conditions' and 'means' acquires a new significance. The 'conditions' are then those material elements of the situation which are not at the agent's disposal in the social sphere, and the 'means' those which he or she does have.

Social relations are, however, also tied to physical elements and their territorial arrangement. Parsons (1952: 90) points out that the orientation of two interacting agents is always related to the organism because the agents' bodies and their personalities, though analytically separable, always form one unit in the concrete action situation. The social aspect of action must therefore always take account in some way of the physical position of the organisms of ego and alter ego: 'All individual actors in social systems are, among other things, physical organisms which must be related in physical space' (Parsons 1960: 276). This means at the same time 'that territorial location[9] inherently enters into all actions' (Parsons 1952: 91). Thus Parsons, in effect, draws our attention to the significance for social reality of the patterns of spatial arrangement of objects in the physical world. Through the consistent integration of the organism of the agent into action theory, the physical context of action can also be included in an analysis of society. Because it is a biological and physical 'object', the agent's body occupies a position in the physical world, and since the agent is tied to his or her body this is 'a crucial fact that always creates a "problem" for action' (Parsons 1952: 90). Since all agents have a body, it cannot be overlooked as a 'focus of orientation'. If the embodied *ego* and *alter ego* are not within sight or sound of each other, they have to use technological aids for communication (e.g. telephones) or change their locations so that they become close.

Apart from these basic general facts, the physical nature of organisms is significant for social relations in two further ways[10]:

> One is that of residential location. The plurality of roles of any individual actor implies a time-allocation between them, and conditions are such that the time-segments cannot be long enough to permit more than limited spatial mobility in the course of the change-over between at least some of them, e.g. family and job. This means that the main 'bases of operations' of the action of an individual must be within a limited territorial area, though 'commuting' by mechanical means has con- siderably extended the range.
>
> (Parsons 1952: 91)

This means that the plurality of roles of every agent structures the course of his or her activities in the physical world. This, together with restrictions on time, gives rise to the fact that the range of actions in the physical world is limited, as they are tied to the locations of the institutions involved (immobile material artifacts). The locations in the physical world which circumscribe agents' activities are referred to by Parsons as 'bases of operations', and the group of persons who meet there at the same time for inter-action is termed 'community' (Parsons 1952: 91). The 'behavioural space' of the physical world is thus structured by the 'bases of operations', which in turn become preconditions for the constitution of 'action spaces' in the social world. Since, however, agents go to the bases of operations in their specific roles, the mediating, but not determinant, significance for the social world of the spatial dimension of the physical world becomes clear.

The second area where the physical nature of organisms becomes relevant relates, in Parsons' view, to the use of force on agents, and the political structuring of association by means of spatial categories. The use of physical force, for example in war, can only happen through the organism of the agent: 'it is *necessary* to "get at him" in the place where he is or would like to be' (Parsons 1952: 91). We can include in this context the necessarily spatial nature of the exercise of political power. Since political power can often only be maintained through the threat or actual use of physical force, which implies that political power depends on controlling physical force (means of violence), its organization in spatial categories is for Parsons a central structural feature of any society. These ideas offer an opportunity for reinterpreting *geopolitical research* in the light of action theory.

In 'The principal structures of community' (1960) Parsons analyses the above ideas in detail, examining precisely the signi-ficance for agents of various territorial locations. He starts by stating 'that a territorial location is always significant as a "place where" something socially significant has happened or may be expected to happen' (Parsons 1960: 251). That is to say, locations in the physical world are only relevant because of their social significance. Parsons classifies the various territorial locations according to their position on a scale of 'bottom' to 'top'. The lowest unit is the 'face-to-face' territory of interaction: the physical area limited by the range of immediate sight and sound. The uppermost unit is the territorial area of validity of national authority.

SPATIAL PATTERNS AND INTERSUBJECTIVE UNDERSTANDING

I now return to the model of intersubjective understanding. Advocates of this model of action, e.g. Berger and Luckmann, conclude *via* their interpretation of Schutz that the everyday world of action is indeed spatially and temporally structured, but that 'its spatial structure is somewhat peripheral to our theory' (Berger and Luckmann 1966: 34). We might deduce from this that the spatial dimension of the physical world has no particular relevance for acting subjects' 'construction of reality', and that therefore it is not worth attempting to identify aspects of Schutz's work which could be integrated into an action-oriented social geography. Yet in the following analysis I shall be demonstrating that, for Schutz, the subject's physical, bodily experiences depend on his or her position in the physical world, and that the spatial dimension is therefore always critical in constituting and understanding the construction of the social world. Particular attention should be paid here to the patterns of spatial arrangement of bodily locations, physical action spaces, and the arrangements of immobile material artifacts. Schutz is interested in two aspects of the spatial dimension of the physical world: how the agent's everyday world is spatially structured, and what consequences this has for subjective processes in constituting social reality.

The basic spatial structure of the everyday world

In something akin to a declaration of principle, Schutz (1962: 222) states that

> man within the natural attitude is primarily interested in that sector of the world of his everyday life which is within his scope and which is centred in space and time around himself. The place which my body occupies within the world, my actual here, is the starting point from which I take my bearing in space. It is, so to speak, the centre O of my system of coordinates.

The immediate 'here' of the body in the physical world is thus the point of departure for all orientation in the physical world. Orientation dimensions and distances are determined with reference to it, and the other structuring factors of the physical spatial dimension of the agent's world operate in relation to it. Schutz, like Parsons, then further distinguishes between the physical world within immediate reach and that within potential reach. This is, in its most general sense, the structure of the argument with which Schutz reconstructs 'the spatial arrangement of the everyday life-world' (Schutz and Luckmann 1974: 36) from the perspective of acting subjects.

The world 'within immediate reach' includes not only immediately perceptible objects, 'but also objects that can be perceived through attentive advertence' (Schutz and Luckmann 1974: 37). This sphere is, in turn, divided according to the agent's sensory modalities. The sensory modalities divide into the agent's range of touch, hearing and sight. For the agent in the natural attitude these work by means of idealizations. In addition, the agent operates in a primary and secondary 'zone of influence'. The primary zone of influence is that area of the physical world on which the agent can have a direct physical effect. It is the zone of direct manipulation and thus the core of reality in the living world. For it is only this area that offers the 'fundamental test of all reality', namely the experience of resistance (Schutz and Luckmann 1974: 42). This primary zone is the origin of all experience of objects in the natural attitude and the immediate area of experience of the world as it is spatially represented. Areas which the agent has up to now been able to turn into a direct zone of experience and influence, through changing the location of his or her body, therefore acquire a greater measure of certainty owing to the knowledge gained from the reality tests involved. Schutz defines the secondary zone of influence as that part of the physical world which the agent can only affect through the use of technological aids. Owing to technological developments, we have witnessed 'a qualitative leap in the range of experience and an enlargement of their zone of operation. It is a leap indeed, once connected to the line of inventions of bow and arrow, smoke signal, gunpowder, etc.' (Schutz and Luckmann 1974: 44). In the secondary zone of influence a further distinction is made between the *possibility* of the extension of this area in a technologically advanced society, and the *actual* extension of the secondary zone of influence achieved by an agent in daily life.

In sum, for Schutz, the physical world is subjectively more important for the agent than those areas which are only accessible to memory and 'embedded in socially objectivized meaning-contexts determined by language' (Schutz and Luckmann 1974: 36ff.). The core of reality is that area of the world which is within immediate reach, 'the specific reality of everyday life' (Schutz 1962: 223).

The distinction between the primary and secondary zones of influence is important, for it can be used to classify the degrees of direct influence and experience of particular significance in constituting the agent's stock of knowledge. Empirical analyses of the everyday worlds of specific societies should, on the one hand, examine the typical distribution of different subjects' accessibility to the commonly, occasionally, and only very exceptionally available means of extending their secondary zones of influence. On the other, they must also investigate which areas subjects can make into primary zones of influence. It should then be possible to establish the regional spatial conditions for the constitution of subjects' available stocks of knowledge.

As I noted, the areas within *potential reach* are further divided by Schutz into the 'world within reconstructable reach' and the 'world within accessible reach'. The former is that area which at one time was the agent's core of reality and is therefore relatively familiar to him or her. It is the area of the past remembered from the orientation of the *point of origin* of experience, the area which, by means of bodily movements in the physical world within immediate reach, has become that within reconstructable reach. The agent assumes that this world will remain the same, other things being equal, even when he or she is physically no longer part of it. Agents also assume that it will be easier to return to where they have been before than to enter other areas. If these ideas are right, they are of considerable importance for the explanation of courses and orientations of action in both the physical and the social worlds.

The area of the world which is within accessible reach lies, from the temporal point of view, in the future. It is the 'world which was never in the reach (of the agent), but which can be brought within it' (Schutz and Luckmann 1974: 38). The fact that this area is not linked to agents' direct past experiences means that they can only estimate the chances of reaching it. Such areas are 'empirically arranged not only according to subjective degrees of probability but also according to grades of ability that are physical, technological, etc. [The agent's] position in a particular time and society is part of the latter limitation' (Schutz and Luckmann 1974: 39).

This means that the accessibility of physical areas depends, first, on what technological means of overcoming distance are available in a society at a particular time; second, on the access a particular agent has to such means. It also depends on the 'hierarchy' of the agent's plans and on available stock of knowledge. At the same time, areas within accessible reach are familiar to different subjects in different ways: certain areas are well known to them although they have never been there; others hardly or not at all.

The zone within accessible reach is thus 'arranged into sub-levels that leave various prospects of attainability. The changes typically decrease in relation to the increasing spatial, temporal, and social distance of the respective sublevel from the centre of my actually present world' (Schutz and Luckmann 1974: 40). The greater the spatial distance and the more inadequate the available stock of knowledge in relation to certain goals and areas, 'the more uncertain are (the) anticipations of the attainable actuality, until they become entirely empty and unrealizable' (Schutz 1962: 226).

Here we have the first indication that Schutz sees the spatial dimensions of the physical and social worlds of action in a relation-ship of reciprocity. If we accept the idea of Schutz and Luckmann (1974: 41) that immediate physical range is a primary pre-condition for a 'face-to-face' interaction, it is clear that the spatial pattern of movement resulting from bodily motion is an important aspect in structuring of social relations. On the one hand, the spatial dimension 'enters into the differentiation of intimacy and anonymity, of strangeness and familiarity, of *social* proximity and distance (between agents)', whereby a spatial differentiation of social relations takes place. On the other, social differentiation is a central 'aspect of the subjective experience of the spatial organ-ization of the life-world'. In other words, a social differentiation of spatial relations is presupposed, 'a system of spatial arrangements extends across the various strata of the social world' (1974: 40f.).

Another aspect of this integrated view of the physical and social spatial dimensions is the fact that Schutz considers that the formal structure of accessibility and reconstructability applies to both the physical and the social worlds. He thus assumes that there is a reasonable chance of a friendship (social world) being formed and re-formed with a person who is not within immediate physical reach.

The social world and the spatial structure of everyday actions

What are the further consequences of the spatial location of agents' bodies, and the corresponding patterns of movement, for the 'construction of social reality'?[11] The agent in the natural attitude has a stock of knowledge involving a high degree of certainty with regard to everything which was once part of his or her primary zone of influence. This decreases in proportion to the increase in indirect experience. The forms of directness/indirectness which Schutz distinguishes in relation to social reality, or 'the life-form of the Thou-oriented acting self', are comparable to those of the physical world. From the social point of view the relationship can be classified according to 'various levels of proximity, depth, and anonymity in lived experience' (Schutz and Luckmann 1974: 61). The range of variation of experience in the social world extends from the direct encounter with another person with whom one is intimate, to a more or less vague notion of 'humanity in general'.

The problem now is how to describe precisely the structures within which agents' experiences of the social world accumulate, bearing in mind their physical attachment to their bodies. An agent experiences his or her other interaction partner *directly*[12] when the immediate physical ranges of both overlap at a given point in time: 'only under those conditions does the Other appear (to him) in his live corporeality: his body is (for ego) a perceivable and explicable field of expression' (Schutz and Luckmann 1974: 62) and through which ego has access to the conscious life of alter ego. Thus it is only possible in the 'face-to-face' situation for ego to match his or her own stream of consciousness of pure duration to that of alter ego, with true simultaneity: both genuinely 'grow old' together. And since the body, as a component of the physical world, always occupies a position in space, the spatial dimension acquires considerable importance as the starting point for constituting meanings in and of the social world. The mediation of the body is the main means for the spatial differentiation of socio-cultural worlds. Before we examine the 'face-to-face' situation more closely, we must first consider the various forms of reciprocal relationship in more detail.

This direct interaction is basically characterized by the fact that ego concentrates attention and experience on to another ego (alter ego). Ego lives in the Thou-orientation among people whom ego experiences in the most general sense as 'equal with himself', regardless of how well ego knows them. The Thou-orientation can be one-sided or mutual. Schutz calls the latter case the we-orientation. By this, social interaction is constituted. He subdivides the we-orientation into different levels, according to the physical and social directness/indirectness in which it is realized: from the simultaneity of pure 'duration' in close intimacy to extreme anonymity and physical distance between the two relating subjects.

All direct social encounters are characterized by the fact that the partners have in common a spatial area, the particular point in time, and their bodies as a vehicle of expression for meanings in their mutual relations. Thus both have at their disposal an abundance of conscious registrations of knowledge. 'Every phase of ego's inner [i.e. pure] duration is coordinated with a phase of the conscious life of the Other' (Schutz and Luckmann 1974: 66). The extent to which this condition of physical proximity can become closeness in the social world depends on the particular biographical nature of the agents' stock of knowledge, on past situations of direct experience. Social closeness depends on the degree of reciprocity in perspective. For Schutz, the socio-cultural gradations of encounters in the spatial–temporal continuum range from the love act between loving partners to the most superficial and non-committal conversation. The most direct social relationship is thus achieved through the greatest possible overlap of physical and social closeness. It occurs in the context of total synchronization of inner and pure duration, and depth and immediacy of experience. The most oblique form of social relationship figures in encounters between two people who are total strangers to each other.

It is particularly important for Schutz that in the we-relationship, the partners can directly test their knowledge: the knowledge acquired from each other in such situations can be seen as the 'surest'. It is just as important, however, that in the direct we-relationship ego must also be aware of alter ego's attitude, and vice versa. In this way their experiences of each other are coordinated and mutually related, so that a reflection of the self in the experience of the other is a constitutive element in this form of relationship. If such forms of relationship are repeated between the same partners over a long period, as for instance in the parent–child relationship or in other locally persistent communities (communes, villages, urban communities, etc.) within immediate or accessible reach, participants build up similar stocks of knowledge through each other. These can be constantly tested within their mutual relationships, in the same way as knowledge about the shared spatial environment. 'In general, it is thus in the we-relation that the intersubjectivity of the life-world is developed and continually confirmed' (Schutz and Luckmann 1974: 68).

The increasingly anonymous strata of relations in the contemporary social world can only be entered upon via deductions from the agent's own encounters, or *via* the typifications of others known to him or her. This again points to the significance of physical proximity as the starting point of all constitutions of the social world. Schutz (1974, 158–9) does however point out that in spite of the physical directness of the encounter, in patterns of institutionalized action the we-orientation is frequently replaced by the they-orientation, as is often the case for example in encounters between buyer and seller. This appears to indicate that Schutz qualifies his emphasis on the significance of the spatial dimension for concrete processes of association. If the social distance is too great, ego is thus, on the one hand, forced to adopt a he/they-orientation even though his partner is physically available. On the other hand, ego can always adopt this orientation even though a Thou/we-orientation is possible, but not vice versa. The point here is that it is clear from this non-reversibility that the significance of the spatial dimension for the social world has not been qualified.

Ego's indirect relations with the social world are based, according to Schutz, on the he- or they-orientation. The reference point of the relationship is no longer the concrete, directly experienced alter ego, but a projected type, which, with increasing anonymity, becomes more and more lacking in substance and less and less connected with specific expectations. Subjective meaning-contexts are replaced by objective, general social meaning-contexts. Every single gradation of indirectness in social relations becomes more anonymous if it is no longer concerned with the world of contemporaries but with the world of the past. Such relations must always have to do with aspects of the agent's own memory-endowed duration or that of the agent's contemporaries, or with the meanings of artifacts produced by previous generations. At all events the experience of the world of the past, and the nature of social attitudes towards it, always remains indirect and one-sided.

The basic spatial structure of everyday actions also has an effect on the courses of subjects' lives. Schutz's theory is that the specific nature of the background spatial conditions surrounding agents' physical and social worlds, and their stock of knowledge, are always expressed in their specific biographies. Agents' physical dependence on their bodies, and their immediate and potential ranges in both the physical and social worlds, result in the formation of quite different stocks of knowledge and therefore of quite different biographies based on them.

Furthermore, following the earliest 'we-relationships' more and more new ones are added, which lead to further selections of life-patterns according to the nature of the immediate and potential ranges of ego and alter ego in the physical and social worlds. Schutz thus opens the way to an in-depth analysis of regional disparities in individual career opportunities. For as the immediate 'we-relationship' is always tied to a common space, and as it is in this form of relationship that 'the intersubjective character of the life-world originally developed' (Schutz and Luckmann 1974: 97), then the formation of spatially differentiated inter-subjective socio-cultural living worlds of meaning can thereby be hypothetically explained.

In my account of this first kind of social relationship – and its more subtle gradations – we have indicated the basic structure of the mediating link between the physical and social dimensions of the everyday world. However, the ranges of human actions in the physical world are not analysed by Schutz within the framework of action space research as such. They are analysed in relation to their significance for shaping social realities in terms of spatial differentiation. The constitutions of social reality are in large measure subject to the condition of space–time overlaps in the physical world. On the one hand, the physical position of the body becomes a precondition for social relations. On the other, social closeness structures movements in space. Since the most important processes in association are bound up with direct spatial encounters, it is easy to see why socio-cultural worlds usually survive. It is easy to see why regional forms of cultures and societies persist despite considerable interregional contact. The deepest social relationships are bound up with the body as a vehicle of expression, and the body in turn is tied to its position in the physical world. The body as a vehicle of expression for inner, pure, and memory-endowed duration fulfils a key mediating function between the immaterial socio-cultural world and the physical conditions which attend any kind of concrete social intercourse.

The social significance of material artifacts

For Schutz (Schutz and Luckmann 1974: 75ff.), interaction with artifacts represents an anonymous social relationship, although the status of artifacts remains ambivalent. While, on the one hand, he assigns them to the anonymous level of social relations, on the other he emphasizes that they can be interpreted 'as proofs of the conscious life of other beings "like me"' (Schutz and Luckmann 1974: 75): as signs, tokens and the results of actions which can be reduced step by step to the original subjective processes. Without this reduction they would be 'nothing but mere objects in the natural world' (Schutz and Luckmann 1974: 75). Thus relations to artifacts are certainly less anonymous than relations to the 'economic system', 'grammar' and similar realities to which agents relate. In the context of social geography, however, it remains important that Schutz sees actions relating to artifacts as anonymous social relations. He describes this interaction between the producer and the user in this way: 'When I make a tool [or other material artifacts] for others to use, the conscious experiences of these others, brought about through their use of the tool (e.g. their recognition that this object is one "to be used for . . ."), are anticipated in my design in the future perfect tense as the goal of my action' (Schutz 1974: 212). The producer of a material artifact thus perpetuates his or her subjective meaning in that object, while leaving it thereafter to the meaning given to it by its user. At the same time, however, the 'constraining character' of material artifacts persists: to a large extent producers control the interpretations of users, even when they themselves are no longer present and their artifacts have taken their place. The world of artifacts has therefore an intersocial effect, and plays an important role in handing down traditions about intersubjective meanings. If we also consider the distinction between mobile and immobile material artifacts, the significance of immobile artifacts, within everyday physical action spaces, for the constitution of agents' stocks of knowledge becomes obvious. Through their temporal, physical and symbolic persistence, they become an important factor in the spatial differentiation of the social world. But they are a factor which is always and only constituted and reconstituted through subjective agency, an agency which, in turn, is shaped by that social world. The point is that in putting the emphasis on subjective agency as the force which shapes the world that presents itself to us in spatial terms, we put the emphasis on the force that can change it.

Conclusions

I have analysed certain sociological theories in terms of the epistemological (three world) criteria developed in the earlier chapters.
These criteria were based on a reading of ostensibly opposed
'objective' and 'subjective' perspectives which drew out their
commonalities rather than their differences. On the basis of these
criteria, I have shown that sociology provides us with numerous
starting points for the development of an action-based social geography in a spatial reference frame. It is however only Pareto, with
his decision-theory approach, Schutz and – with reservations –
Parsons whose theories relate systematically to the models of
action corresponding to the three worlds. Other authors are either
inconsistent in their approach to action (Simmel, Durkheim,
Halbwachs), or treat it unsystematically (Weber). This does not
mean these theorists should be dispensed with in an action-
oriented geography. In detailing their general models of action, I
have attempted to draw out their spatial bearings in the context of
social geography. I have argued that it is necessary to identify
spatial frames of reference that correspond to the physical, subjective and social spheres of life so that the risk of reductionism, or
the privileging of any sphere, can be avoided. Far from being a
'cause' of events, space, like material conditions, is a complex
constraint on them. This constraint can only be understood in
terms of the different levels on which it works, and it never 'con-
strains' any level in the same way.

More specifically, the general sociological models of action can
be developed in relation to an action-oriented social geography
through consistent attention to the physical–material conditions
of actions. There are several points of departure for this in the
sociological literature.

The first point in considering the physical–material and biological conditions of action is the acknowledgement of the fact that a large number of actions are realized in the physical, bodily situation. It is for this reason that the spatial dimension acquires particular social importance for 'face-to-face' interactions in particular, whose role in the construction of social reality is stressed especially by Simmel and Schutz.

The second point is acknowledging the importance of the spatial patterns of arrangement of material artifacts for concrete actions. Durkheim, Weber and Schutz also draw attention to the meanings preserved in artifacts. The associations established in this material way are perpetuated in artifacts, which means that artifacts contribute to the physical differentiation of social phenomena.

The third point in the analysis of the importance of the physical world for social actions emerged through an analysis of Pareto, Simmel and Halbwachs. It is that material objects and artifacts which can occupy a clearly definable position in space become vehicles for symbolic encoding. The symbolic relations do not thereby become material facts. But since they only become relevant for agents in connection with particular material objects, such as place-names, symbolic relations to places should also be seen as spatial differentiations of phenomena in the social world.

It is clear that the geopolitical, time-geographical and action-space research approaches in social geography could, within the framework of an action-oriented reinterpretation, be critically synthesized. This is obvious against the background of Parsons' and Schutz's action theories especially. They indicate what social significance the constraints resulting from spatial patterns of arrangement, and the spatial patterns of the movement of organisms, have for the lives of acting subjects.

In this book I have argued that primacy has to be given to subjective agency in social geography, at the same time as social geography has to take constant account of the physical–material conditions surrounding human action. In social geography thus far, the traditional areas of research have been the investigation of objective patterns of spatial arrangement as such, the activity involved in their production and use, and the relationship between 'man and his (socio-cultural) environment'. None of this should be excluded from an action-oriented geography. But in addition to these areas of research, the examination of the effects of objective patterns of spatial arrangement on human actions, meaning the social significance of the patterns in spatial arrangements of material artifacts, and of positions of agents' bodies, should be integrated into research in social geography.

I shall briefly recapitulate some other, earlier arguments that led to this conclusion. As we saw in Chapter 2, the procedure of situational analysis proposed by Popper for the social sciences can be understood as a more precise expression of the call to implement the 'biological approach to the third world' (1979: 112). Just as, according to Popper (1979: 112), a biologist may be interested in both the behaviour of animals (or spiders) and the inanimate structures produced by animals (spiders' webs), so there is a dual area of research interest for social scientists. On the one hand, there are human actions which, under particular situational conditions, produce or make use of a particular pattern of spatial arrangement, and, on the other, there are the intended and unintended objective consequences of actions in spatial arrangements.

In Popper's theory these two areas of investigation (acts of production and objective consequences as such) are kept separate. The study of acts of production should not be 'infected' by the subjective perspective. The action situation as it appears to the agent might be of interest, but that, subjectively speaking, is all. In situational analysis the deductive method of explanation is used, as it is by agents themselves. The central question here was: which objective items of knowledge, thoughts and ideas 'as such' lead to provisionally verified or conclusively falsified courses of action? The fact that subjective states of mind, world 2, play only a subordinate role here is justified by Popper on the ground that these states can be logically derived 'in almost every respect' from world 3.

Popper's 'objective consequences of actions' in relation to human geography means that the pattern of spatial arrangement of the material artifacts of a particular society – especially the geometric form and recurrence of these patterns and their evolutionary changes in the context of scientific progress by means of the falsification principle – are paramount.

Popper's procedure thus has as its aim the explanation of human actions on the basis of their objective structure. He isolates two important questions. First, there is the question of the possible effects of the configuration of the objective structure upon agents' actions. These must be systematized as the situational factors within range of the agent at the macro or micro level. Micro- and macro-analytical areas could be logically and consistently related to each other. Second, there is the question of the respective functions, in particular courses of action, of the separate elements of the objective structure of (spatial) arrangements.

It should be clear from the above that the approach centred on the *course* of the action, which systematizes action in particular situations, and the approach centred on objective structure, can sometimes be integrated. They are not, inevitably, mutually exclusive. From the structural standpoint, it is just as important to know what consequences the pattern of spatial arrangement has for an agent in a particular situation, as it is to know alone the consequences of agent's actions. In both cases, however, the material should always be analysed from the objective point of view, in the Popperian sense, through the application of the deductive method of explanation (situational analysis) according to the falsification principle (rational explanation). There are clearly similarities here with the spatial–scientific methodology. There are also important differences.

Within the framework of an action-oriented social geography, the objective spatial structure of various typical, regularly occurring, goals of actions should be differentiated. It is important in such an analysis to ascertain which spatially represented elements are integrated into courses of action, so that, in the case of problem situations, appropriate technologies can be formulated according to Popper's procedure of situational analysis. Thus by means of a sociological methodology, existing action-oriented social theories could be expanded by including the physical–material components of human existence, without inappropriate reductions and hypostatizations. To avoid these, schemes of reference for the location and understanding of socio-cultural factors would have to sit beside any analysis of the physical–material aspect of the action situation.

Accordingly, general hypotheses concerning regularities of action, and regularities in the elements of objective structures have to be set out for social research in human geography. For 'goals', 'means' and 'consequences' of actions, distinctions should be made between:

- 'goals' in the socio-cultural, subjective and physical–material worlds;

- 'means' in the socio-cultural, subjective and physical–material worlds;
- 'consequences' in the socio-cultural, subjective and physical– material worlds.

If these preconditions can be met, situational analysis could become a fruitful procedure in social geography. Appropriate suggestions could then be made for the solution of various problem situations. We would also be in a position to release geography from its spatial bondage, enabling it to become an objective social science where actions in space, rather than space as such, is the object of investigation.

Let me now review my conclusions concerning phenomenology, and Schutz's epistemology. Two aspects of his concept of empirical research seem particularly significant in relation to social geography. These are: (i) the fact that the study of actions and that of artifacts have the same status from the methodological point of view; (ii) Schutz's insistence that in order to comply with the postulate of causal adequacy, the subjective meaning of actions must be included in the investigation.

In Schutz's model of intersubjective understanding, it is methodologically immaterial whether an action leads to a completed concrete result (a gesture, a bodily movement) or to an artifact. Two topics of research are hereby opened up here for an action-oriented geography from a subjective perspective: the actions of members of a society under particular physical–material conditions on the one hand, and mobile and immobile artifacts in their spatial arrangement and social significance for actions on the other.

In fact, both areas of research can be approached either from the objective or the subjective point of view. In the objective approach, the aim would be to investigate the relevant facts 'as such'. They are separable from the subjective processes whereby the meanings of facts are constituted. The objective approach then aims to deduce production processes from the structures and the patterns of arrangements, without taking account of the subjective meanings of the actions which produced the artifacts. However, according to Schutz this procedure should only be a *pro tem* one. It should be abandoned once successful explanations are found through the investigation of the intersubjective meaning-context. In this subjective perspective, the pattern of spatial arrangement of artifacts is seen as a reference area of human actions in which the subjective meaning of their producers acquires quasi-permanence. It is necessary to explain these facts by 'tracing them back to meaning-endowment by a conscious mind' (Schutz 1974: 193).

This requirement bears especially on the construction of material ideal types (models), and also on the typology of courses of action. For Schutz, the anonymous world of 'contemporaries in the past' must be gradually brought back from absolute distance to the absolute closeness of experience. Typical action and typical 'in-order-to' motives of agents must be deduced from the typical patterns of spatial distribution of artifacts. On the other hand, we can also deduce typical actions, or typically produced artifacts, from a personal standpoint. Thus, a dual strategy is opened by a subjective methodology. First, an attempt can be made to construct personal ideal types, and then the expected artifact or the expected pattern of spatial arrangement of material artifacts can be deduced from these. Both can be compared with the personal ideal type, and agreements, closeness and deviations established. Second, one could construct material ideal types or models and deduce from these the previous or past subjective meanings of the action, and so of the agent.

In both subjective strategies, the form of spatial arrangement of the 'finished artifact', as Schutz calls it, should be seen as evidence of past subjective meanings of actions in an earlier contemporary world. But we can also 'fantasize', on the basis of personal ideal types, about spatial arrangements in the future, and try to formulate prognoses. Such predictions would contain no element of certainty, however. Even if 'in-order-to' and 'because' motives are currently relevant and adequate, they can be proved wrong subsequently. The future is and remains absolutely undetermined and indeterminable. Attempts to explain an action or an artifact, or pattern of spatial arrangement of material artifacts, by referring to an intention or goal, will fail if, after empirical testing of the hypothetical explanation, it becomes clear that the result of the action does *not* coincide with the intention. We would fall victim to Popper's 'conspiracy theory of society' if, in spite of empirical evidence to the contrary, we described an unintended result as an intended consequence of the action.

It should now be clear that in an action-oriented geography, the objective and subjective perspectives are not mutually exclusive. I have already shown that the two perspectives actually complement each other. In problem cases where the meaning-structure can be established without including the subjective perspective, and in those cases where the researcher knows (intersubjectively) the meanings and consequences of the actions under investigation, the objective perspective is enough. If the explanations prove empirically untenable, and the actions suggested do not have the expected results, the subjective perspective should then be adopted.

However both the objective and subjective epistemological standpoints are in themselves too general to be used in the development of a social theory for geography. For this, we need the action theories of sociology, which I reviewed in the first part of this conclusion. But what emerged from the analysis of these theories was that, unless the subjective perspective is borne in mind in the sociological analysis of action, we are without the means to counter the assumptions about spatial causalities which still shape so-called human geography. I tried to synthesize these aspects of the various sociological action theories which point to a different system of explanation, a system which recognizes that only subjective agency, however constrained, could move the structures that constrain it. Yet, in doing so, I encountered a persistent theoretical constraint: subjective agency was always assigned, paradoxically, to the periphery of action. The fact that subjective agency is trying to return to centre stage in some worthy post-structuralist evocations that are ignorant of its marginalized history does not help its progress. Yet it needs to return to centre stage, if the territorial, ecological and social discriminations of racism and sexism are to be formulated as soluble problems. Of course, whether they will be solved is another question.

Appendix

Risky concrete research proposals

This appendix is an attempt to indicate some key areas of research for an action-oriented geography. These might encompass:

- the development of appropriate spatial frames of reference for every ontologically different area of reference, within both the objective and the subjective perspectives of research;

- devising rules which facilitate comparisons between the simultaneous locations of a material artifact in the physical and the social worlds;

- the investigation of the significance of closeness/distance in the physical world in initiating and maintaining social relations;

- the investigation of the significance of the physical world aggregation of immobile artifacts, as the (intended/ unintended) consequences of past actions, for the present social world and relations in it;

- the investigation of the spatial structuration of courses of action around immobile artifacts *and* their significance for forms of interaction in the social world;

- the investigation of the meanings preserved in immobile artifacts *and* of the significance of such meanings as social conditions, mediated through material things, for particular types of action within certain spatially delimitable areas of action;

- the investigation of the significance of the social contents of artifacts as 'forms of association' and as frames of reference for action orientation;

- the investigation of the significance of the symbolic contents of places and place-names for action orientations, decisions, courses of action and the initiating of social relations outside the agent's own place of abode;

- the investigation of patterns of spatial arrangement as the intended/unintended consequences of actions and as the occasion, furtherance or constraint of further actions;

- the investigation of the role-specific differentiation of (physical world) action spaces and their significance for the social context of action;

- the investigation of different societies with regard to their prevailing typical distribution of various subjects' access to common, occasional or only exceptionally available opportunities to extend their secondary zones of influence;

- the investigation of those areas which subjects can regularly make their primary zone of influence. The systematic study of these two topic areas will help in the understanding of the significance of regional conditions;

- the investigation of the significance of action spaces in the physical world for regional disparities in individual career opportunities;

- the investigation of meaning-constitutions of social reality based on regionally differentiated stocks of knowledge, and of the development of concepts of communication between different socio-cultural worlds.

Notes

1 Space and causality, or Whatever happened to the subject?

1 The philosophical debates are summarized in Smart (1964), Alexander (1956), Nerlich (1976), Sklar (1974) and Mellor (1981).

2 Geography, unlike many disciplines, at least gives public credit to one founding mother: Ellen Semple.

3 Kant, *Physische Geographie*, Königsberg, 1802.

4 Lefebvre, 1974 (1981): 151–2.

5 It is true that 'behaviour' is used by many social scientists and geographers as an umbrella term for all human activities. I do not consider such an application as useful at all. Apart from inadequacies which I shall examine later, it leads either to substantial loss of information and considerable confusion, if we are not told in which sense 'behaviour' is meant, or to tedious elaboration if we are. I suggest 'activity' as an umbrella term and see 'action' and 'behaviour' as two specific models representing human activity.

6 In referring to 'action' I mean 'social action', as postulated by Max Weber (1968: 1). That is to say, a human activity oriented towards 'doing' or towards the results of human activity (artifacts). It may also be an activity that affects nature, altering through socially derived knowledge physical objects which have not as yet changed by human hands. Since with this theoretical approach there is no action which does not at the same time imply social action (see Övermann *et al.* 1979), I shall use the shorter term 'action'.

7 See Wright's (1971: 84–5, 58) distinction between 'quasi-causal' (causal description of intentional acts) and 'quasi-teleological' (intentional description of causal processes, in the sense of functional explanations) and Luhmann (1962: 618).

8 'Project of action' (Weber 1968) is referred to in the literature by various names. Habermas uses the term 'act preparing for argumentation', Wright uses 'inner aspect of the action' or 'forming on the intentionality of action', and Schutz 'planning in relation to motives'.

9 The distinction between 'end' and 'means' with regard to an action is not simple. For Prewo (1979) this is a reason to talk about 'the

labyrinth of the concept of action'. We can take as a principle that anything undertaken or included in the action in order to achieve a goal should be described as 'means'. But the goals, the project of action, can extend to varying degrees. They may range from an aim in life to plans for a single act lasting only seconds or minutes, and they may correlate to each other to a greater or lesser extent. If, for instance, an objective for the day helps to achieve a goal set for a year, the former becomes a 'means' in the light of the latter, and so on. A clear distinction between 'end' and 'means' is therefore only possible in relation to the reach of the action's project.

10 This typology of empirically occurring actions and social relations does not claim to reproduce every possible form of social process or event. It should be understood merely as an illustration of action-theory approach to social reality.

11 I shall be dealing with these implicit basic assumptions in more detail in the following two chapters.

12 When Mead's 'symbolic interaction' (1967) seeks to overcome this dilemma by means of a distinction between 'Me', 'I' and 'Self', it nevertheless remains incomplete from a sociological point of view since it adheres rigidly to the behaviourist position, retaining un-critically the stimulus–reaction pattern. Mead's 'symbolic interaction' thus offers no appropriate access to the investigation of social relations, for it has to presuppose the social, like all other research concepts with a basic behaviourist substructure.

13 Mischel's (1981) critical approach is from another, although related, angle. He points out that the key concepts of cognitive behaviourism (desire, motive, level of expectation, etc.) are all basically intentional in character. That is, cognitive behaviourism implicitly describes human activity in terms of intention. Mischel advocates that this fact should be recognized and, as a consequence, 'action' and not 'behaviour' should be the term used.

14 For relatively recent examples of this approach, see Gold (1980) and Walmsley and Lewis (1984).

2 The objective perspective

1 Popper is not consistent in his distinction between 'commonsense' and 'common sense'. I have also used these terms interchangeably in this chapter.

2 Popper (1979; 38) sees the scientific positions of naive empiricism, logical positivism, phenomenalism and phenomenology, and simi-larly, all behaviourist approaches as contaminated by the mistaken commonsense theory of knowledge.

3 As far as the temporal aspect of perception on the one hand is concerned, and that of the dispositions, horizon of expectations and hypotheses on the other, we can admit with Popper (1979; 345) that a new hypothesis will generally be preceded in time by those obser-vations which were directed by the earlier horizon of expectation.

Yet this must not be understood as saying that observations generally precede expectations or hypotheses. On the contrary, each observation is preceded by expectations or hypotheses; by those expectations, more especially, which make up the horizon of expectations that lends those observations their significance; only in this way do they attain the status of real observations.

(Popper 1979: 345–6)

The first horizon of expectations in connection with possible actions is, according to Popper, present at birth (1979: 347).

4 Popper refers to Tarski's concept of truth (1969: 261), 'that a theory is true if and only if it corresponds to the facts' (Popper 1979: 44; see also 326ff.), and applies it to the realism postulate of common sense. On the other hand, we also see here Popper's specific interpretation of the aspect of expectation, of the intention component of an action. It is not a question here of the goal-directed, finalistic teleology propounded by economist Marxism and other historicist approaches, as Popper calls them, but of goal orientation, which has to be constantly tested with the possibility of failure, i.e. falsification. It is possible, in other words, for intentions to founder.

5 For Popper, the results of scientific actions are of immense importance in relation to commonsense actions. Very often these results help to dispel commonsense theories (e.g. the theory that the earth is flat), ensuring that they are replaced with other theories 'which may appear to some people for a shorter or longer period of time as being more or less "crazy"' (1979: 34) (e.g. the theory that the earth is round). The assumptions of common sense are often changed in this way. However, 'if such a theory needs much training to be understood, it may even fail for ever to be absorbed by common sense' (1979: 34).

6 The argument is as follows: world 3 is only possible because of the descriptive and argumentative or critical function of language, and it is only human language which has access to this function. Animals only have access to the two lower functions of human language: self-expression and signalling, which make up the precondition for the two higher functions (see Popper 1979: 119). We should also point out with Bühler (1976: 90ff.) that it is the descriptive function of language which makes empirical research possible, and the argumentative function which is a precondition for any criticism.

7 To illustrate the independence of world 3, Popper describes two hypothetical experiments (1979: 107–8):

Experiment (1). All our machines and tools are destroyed, and all our subjective learning, including our subjective knowledge of machines and tools, and how to use them. But *libraries* and our capacity to learn from them survive. Clearly, after much suffering, our world may get going again. Experiment (2). [In addition to (1)] all the libraries are destroyed also, so that our capacity to learn from books becomes useless. If you think about these two experiments, the reality, significance, and degree of autonomy of the third world (as well as its effects on the second and first worlds) may perhaps becomes a little clearer to you. For in the second case there will be no re-emergence of our civilization for many millennia

(Popper 1979: 108)

In connection with the questions asked by Husserl and Schutz (see Chapter 3), we can formulate a third experiment, though it shows Popper's idea to be less comprehensive. Not only are our technology and libraries destroyed, but so is our capacity to remember and transmit the information we have learned. This does not mean that world 3 is not independent. It does mean that the social importance of world 3 (or at least part of it) depends on its entering the consciousness of social agents.

8 Quoted by Popper 1979: 109.
9 Of course the famous opposing view is that of Kuhn (1970).
10 Such hypothetical statements can be ordered 'thematically' when they are concerned with the same basic area, the same object of knowledge. 'Logically' systematized means that a number of statements form a deductive system, i.e. a system of propositions whose area of meaning is arranged hierarchically. From propositions with a large element of generality, all more specific propositions can be logically and deductively derived. All specific propositions with little information content are subordinate to a hypothesis which has greater general validity.
11 See especially Popper 1979: 106–90 and 206–55.
12 The differentiation into three worlds on the basis of the ontological status of the various facts of reality is of particular importance for the development of an action-theory social geography. It will be referred to in Chapter 6 in connection with the space problem.
13 Popper's original reads: 'la rationalité comme attitude personelle consiste dans la disposition à corriger nos idées. Dans sa forme la plus développée, intellectuellement, c'est une disposition à examiner nos idées dans un esprit critique, et à les réviser à la lumière d'une discussion critique avec autrui' (Popper 1967: 149).
14 This aspect of Popper's doctrine of method has already been dealt with in detail in numerous introductions to scientific theory (see Esser *et al.* 1977; Prim and Tilmann 1979; Opp 1970; Stegmüller 1975; Kutschera 1972, and others). I introduce it briefly here in the most general context for two reasons: first, in order to be able to refer to it in the context of Popper's critics; and, second, so that I can assess more precisely the deviations and convergences of the methodologies of Popper and Schutz.
15 Popper (1979; 1982) uses the term 'explicandum'. I shall use the more familiar term 'explanandum'. Popper (1979; 1982) uses the term 'explicans' for 'explanans'.
16 Popper occasionally uses the expression 'situational logic' in place of 'situational analysis'. In his more recent publications, however, he prefers to use the term 'situational analysis', since the other 'may be

felt to suggest a deterministic theory of human action; it is of course far from my intention to suggest anything like this' (Popper 1979: 178).

17 See Jarvie 1972; Lichtman 1965; Koertge 1974, 1975, 1979; Watkins 1972; Donagan 1975; Brodbeck 1968; Schmid 1979.
18 Actually, spiders seem to have a peculiar fascination for certain philosophers. Marx also refers to spiders and webs in another context. But this is to digress.
19 Popper's general ideas on the actions situation have already been examined in detail, so I shall concentrate here on the methodological aspects.
20 The original reads: 'que les personnes ou agents . . . agissent de façon *adéquate* ou *appropriée*, c'est-à-dire conformément à la situation envisagée.'
21 'Bien entendu un principe à peu près vide.'
22 In the original, 'Ce principe ne joue pas le rôle d'une théorie empirique explicative, ou d'une hypothèse testable'.
23 See Weber 1951: 190ff.; 1968: 19ff.
24 For the causal position see Albert 1964, 1972; for the criticisms on Popper's 'causal' methodology see Habermas 1984, 1987, 1988 and Wright 1971.
25 Originally, 'le seul moyen que nous possédions pour expliquer et comprendre les événements sociaux'.
26 See Jarvie 1972: 5ff.
27 See in this connection Koertge (1979: 91ff.) who leaves out of consideration the component of social engineering.
28 See Popper 1979: 217ff.

3 The subjective standpoint

1 We cannot examine here all the imprecisions and contradictions inherent in this approach. The main difficulty is that Husserl himself was continually criticizing and questioning earlier phases of his own work, so that it is impossible in the present study to give a consistent account of his line of thought. The various introductions to Husserl's phenomenology, their criticisms and applications in sociology and history, usually concentrate on a particular phase of Husserl's work. It would therefore be necessary first to compare the secondary literature with Husserl's own self-criticism, in order to assess the adequacy and correctness of the former (see Adorno 1982; Esser *et al.* 1977: 84–96; and Rombach 1974). Critics moreover do not agree on the delimitation of the different phases and stages of Husserl's work, which makes classification of the critical works extremely difficult (see Szilasi 1959; Diemer 1956; Held 1981).

It is thus no simple matter to provide a consistent account of Husserl's ideas. Since this section turns mainly on the basic premises of the 'subjective perspective' of social science as they concern knowledge theory and scientific theory, in the following analysis I shall be concentrating on those aspects which, in discussion with Husserl,

Schutz mentions as necessary for the founding of a 'non-realistic-objective social science' as Schutz calls it in an interview with Baeyer (1971: 10).

2 For the importance of logic and mathematics for scientific action in the phenomenological theory of knowledge, see Szilasi 1959: 29ff.

3 'Über den Begriff der Zahlen' (1887) and – resulting from a highly controversial discussion with Frege – 'Logische Untersuchungen' (1901).

4 See Peursen 1969: 25; Pivčević 1970: 93 and Held 1981: 278.

5 In the 'artificial attitude', i.e. in its most general sense the attitude of philosophical doubt, it should be possible to establish which aspects of the perceived object have been constituted by the subject's act of cognition, and which are attributes of the object itself.

6 See Chapter 5.

7 Equating Husserl's 'life-world' with Schutz's 'everyday world' is not without its difficulties, since Schutz's retains the term 'life-world' as a central theoretical element (see Schutz and Luckmann 1974). Grathoff (1978b: 68ff.) suggests that we can begin to distinguish the two terms by interpreting Husserl's 'life-world' as 'the background to the everyday world' (1978b: 69): as a philosophical classification and prestructuring of the presuppositions at the basis of all action, which enables us to describe the everday world in non-everyday, theoretical language. It makes possible a scientific description of the structures of everyday life and the natural attitude. According to Grathoff (1978b: 68) we should interpret Schutz's 'everyday world' as the

social construct, always perigee, of a world already manifoldly pre-constituted in its history which is always concrete. 'Everyday life' (on the other hand, is) the possible constructive creation of a new world out of everyday difficulties, through constant action and inter-subjective experience – together with the resulting insight into its possible obsolescence and destruction.

(See also 1978b: 73ff.)

Schutz is particularly concerned to show that the agent's life-world is from the first not his own private world but an intersubjective world: the 'fundamental structure of its reality is that it is shared by us' (Schutz and Luckmann 1974: 4). This common element has reference to both the socio-cultural and the physical-natural sphere of the life-world, and thus to everything which exists there. Schutz also stresses that nature is not perigee as an intersubjective common element because

the significance of this 'natural world' – which was already experienced, mastered, and named by our predecessors – is fundamentally the same for my fellow-men as for me, since it is brought into a common frame for interpretation. In this sense, the province of things belonging in the outer world is also social for [the agent].

(Schutz and Luckmann 1974: 5)

8 I should like first to correct a misconception which appears to be gaining ground in the specialist literature of human geography. For many scholars phenomenology is a method 'which seeks to understand the life-world of man directly through "holistic" interpretations of everyday situations' (Seiffert 1972: 17). In Seiffert's view a phenomenological explanation is one where 'on seeing a formulation and interpretation of experiences and feelings, we have a "yes-that's-how-it-is experience"' (1972: 29). Interpretations in human geography amount to the idea that 'phenomenology' provides welcome metatheoretical grounds for raising 'common-sense', with all its personal prejudices and vague terminology, to the rank of 'science' (see Wirth 1979 and for further critique of this attitude Pickles 1985). Moreover, the attempts of phenomenologist philosophy to show that every non-alienated science must have its foundations in the life-world should not be misconstrued to mean that science should be satisfied with formulating subjective–descriptive statements in everyday terms about the everyday world. Reference to the importance of the life-world for science should not be misused to justify the scientist's own uncontrolled and uncontrollable scientific action or a lack of scientific rigour. Scientific understanding of the 'subjective perspective' demands a scientific method appropriate to the life-world, but not the raising of subjective and doctrinaire statements to the rank of science. The 'life-world' is turned into a scientific problem, and the natural, or naive, attitude is not held to be the only possible access to the world in which we live. In the following I shall attempt to show that, according to the phenomenology of Husserl and Schutz, all science has its roots, or starting point, in the life-world. This does not mean, however, that science is impossible.

9 Husserl (1982: 128–9) gives the following more precise definition of the term 'reality':

> In a certain way, and with some caution in the use of words, we can also say that *all real unities are 'unities of sense' [Einheiten des Sinnes]. [. . .] An absolute reality is just as much as a round square.* Reality and world are names here precisely for certain valid *unities of sense* [*Sinneseinheiten*], unities of 'sense' [*Einheiten des Sinnes*] related to certain concatenations of absolute, of pure consciousness which, by virtue of their *essence*, bestow sense [*sinngebende*] and demonstrate sense-validity [*Sinnesgültigkeit*] precisely thus and not otherwise.

10 The reality, namely, which is perigee before all other realities and which, although existing independently of the self, does not have independent meaning-content.

11 See Lüchinger 1982: 36.

12 By 'apperception' Schutz means the process through which concrete experience or perception is interpreted with regard to the stock of knowledge.

13 Every expectation that an experience will conform to type can be confirmed or refuted in the actual process of experience.

If confirmed, the content of the anticipated type will be changed; at the same time the type will be split up into sub-types; on the other hand the concrete real object will prove to have its individual characteristics, which, nevertheless, have a form of typicality.

(Schutz 1962: 8)

It is, however, often the case that the agent in fact places more importance on the non-typical aspects of a 'specimen' of a type, e.g. 'cat'. It is not what my cat has in common with other cats that I find most important – on the contrary, the important thing is what distinguishes her from all the other cats I know. In other words, I may perceive an object as a specimen of a type, but I do not have to think of it in terms of general categories.

14 A somewhat more detailed definition of the 'biographical situation' is that the agent 'finds himself in a physical and socio-cultural environment [. . .], within which he has his position, not merely his position in terms of physical space and outer time or of his status and role within the social system but also his moral and ideological position' (Schutz 1962: 9).

15 See Schutz 1962: 230ff.

16 The scientist should be 'disinterested' insofar as he is not motivated by the same (practical) intentions as the agents whose actions he is investigating. He thus takes the 'leap' (in Kierkegaard's sense) into the 'theoretical' attitude, into a different 'attention à la vie'. 'The theoretical thinker once having performed the "leap" into the disinterested attitude is free from the fundamental anxiety and free from all the hopes and fears arising from it' (Schutz 1962: 247).

17 'Problems as such' do not have the same status for Schutz that 'truths, ideas and problems as such' have for Popper; see Chapter 4 in this context.

18 There is a certain similarity between this and Popper's postulate of a world 3 existing independently of the knowing subject. But, as opposed to Popper, Schutz believes that it is not the knowing subject who is independent of objective ideas, but the scientist who in the theoretical attitude is independent of the concrete outer world. Thus in Schutz's view theories can be 'dropped' without human beings being affected: the scientist suspends his concrete self within the framework of the 'theoretical attitude'.

19 It should not, however, be inferred from the following that Schutz sees the methods of the social sciences and those of the natural sciences as widely divergent. He rejects as untenable the postulates both of the nomothetic school of sociology, i.e. sociologists who follow the example of the natural sciences and are therefore only interested in the formulation of rules and the explanation of phenomena, and also those of the idiographic school which is only interested in unique statements and in the understanding of phenomena, spurning measurement and experiment in sociology. According to Schutz (1962: 48ff.), the latter in particular overlook the fact that certain procedural rules are equally valid for all the empirical

sciences: the principles of controlled inference; the principles of controlled verification by other scientists; the theoretical ideals of unity, simplicity, generality and exactness. Schutz considers these common factors valid, whatever the object of research. For the moment I would just like to point out that Schutz proposes a third orientation of sociological methodology which should not be equated with anything previously propounded.

20 Including the reference to the value-free postulate.

21 See further analyses in Chapters 6 and 7.

22 By 'action' Schutz means here only those activities which are directed towards the external world.

23 See Schutz and Luckmann 1974: 99; Habermas 1984: 139ff.

24 Schutz gives very complex definitions of the situation during the course of his theoretical works on action theory in sociology, but this outline will suffice here.

25 See Chapter 2.

26 See Popper's theories on world 3 in Chapter 2.

27 Both Weber and Schutz mean by 'motive' not a cognitive element as it is understood in motivation psychology, but a meaningful *reason* for action (see Schutz 1972: 86 and 229). I shall be dealing in more detail with the difference between 'in-order-to' and 'because' motives in Chapter 5.

28 In answering this question Schutz refers to the necessity for a clear distinction between 'genuine' and 'pseudo genuine' 'because' motives. He points out that every 'in-order-to' motive can be restated as a 'because' motive but not vice versa. All reformulated 'in-order-to' motives must be described as 'pseudo genuine' 'because' propositions, since they do not take account of the temporal nature of the reason for the action.

29 Schutz agrees with Weber that both kinds of motive are important for the realization of this objective. Weber's statement that it is the task of sociology to 'interpret social action ['in-order-to' motive] and thereby explain the causes of the action and its effects [because-aspect]' (1968: 1), refers implicitly to both motives. Methodologically, however, he does not give equal importance to both, and the interpretations of methodologists of the causalist school have been somewhat one-sided. For Schutz (see 1964: 13), social phenomena only make sense if their 'in-order-to' or 'because' motives are shown. Explanations of human actions involving reference to these two types of motive are compatible but not identical. The reason for this is that *what* they explain and *how* they explain actions are two radically different things. Because they refer to different aspects of actions they are more complementary than excluding modes of social explanation. In Chapter 4 I shall give a more general account of this.

30 The abundance of clues given directly to the observer in the face-to-face situation – to guide him in his interpretations – are only indirectly available in the world of contemporaries. But here also there are varying degrees of indirectness. Understanding the world of contemporaries, which once or several times previously was part of the

observer's *own* face-to-face situation, is relatively straightforward. This situation in the world of contemporaries can be seen as a variation of the face-to-face situation, since it can be shown to be a variant function of the knowledge at hand which the observer has regarding the situation. On the other hand, 'the constitution [of the] contemporary world out of *other's* past face-to-face experiences is at the moment of revelation not only his, but also [the observer's] contemporary world' (Schutz 1974: 254). In this case the observer is thus dependent on the written or spoken reports of others – a situation with which all consumers and producers of regional studies are familiar. Similarly – and this is the third degree of indirectness – every conclusion regarding the intention of the producer of an artifact is an indirect conclusion within the process of understanding. For it is an object without consciousness, one which conveys to the interpreter signs of another's intentions.

31 Schutz differentiates here between social relations in space and those in time. If agents directly relate to each other both spatially and temporally in a mutual 'Thou-orientation', Schutz speaks of *direct interaction*. If this 'spatio-temporal community' (1972: 142) remains intact, but there is only a one-sided 'Thou-orientation', he uses the term 'face-to-face relationship', or 'face-to-face situation'. If there is 'temporal community' between agents but no 'spatial community', the world of directly experienced social reality [*soziale Umwelt*] becomes a world of (mere) contemporaries [*soziale Mitwelt*]. In this situation the agents enter into a 'we-relationship' (1972: 180), although there is no direct experience of the person whose actions are to be interpreted.

32 See Chapter 2, where Popper describes the rationality principle as a 'nil hypothesis'. As far as the history of the discipline is concerned, it is important to remember that Schutz introduced this idea into sociological methodology about thirty years before Popper.

33 See Chapter 2, and Popper's inverse proportional relation of the decreasing information-content of a hypothesis with the increase in precision of the 'if-components' on the one hand, and of the increasing information-content with the increase in precision of the 'then-components' on the other. This can be seen as a further congruence of the two perspectives.

34 Although there is no inherent meaning as such in artifacts, they should still be seen in relation to the 'in-order-to' element present in the original action context.

35 In Popper's language we could thus say that Schutz's scientific model is also characterized by a deductive methodological principle. Neither the everyday agent nor the scientist operates with inductive generalizations. Where they differ is in their prime interests: the one is putting to the test his stock of knowledge constituted in face-to-face situations, the other the theory prevalent in his discipline.

36 Weber (1968: 11–12) defines a 'meaning-adequate' and a 'causal-adequate' understanding as follows.

The interpretation of a coherent course of conduct is 'subjectively adequate' (or 'adequate on the level of meaning'), insofar as, according to our habitual modes of thought and feeling, its component parts taken in their mutual relation are recognized to constitute a 'typical' complex of meaning. It is more common to say 'correctly'. The interpretation of a sequence of events will on the other hand be called *causally* adequate insofar as, according to established generalizations from experience, there is a probability that it will always actually occur in the same way. [. . .]. Thus causal explanation depends on being able to determine that there is a probability, which in the rare ideal cases can be numerically stated, but is always in some sense calculable, that a given observable event (overt or subjective) will be followed or accompanied by another event.

37 See the relationship between 'in-order-to' and 'because' motives and their importance in action orientation, in Chapter 5.

4 An epistemological synthesis

1 See Buttimer 1974, 1976; Ley 1977, 1978; Ley and Samuels 1978.
2 See Bernstein 1976; Habermas 1984, 1987.
3 See Galtung 1977: 24ff.

5 Social theories of action

1 See Bubner 1982: 66–89. Bubner is first concerned to make a clear distinction between the terms 'work' (production) and 'action' (practice), which features in the argument between Aristotle and the Platonic–Socratic philosophy. The terms 'work' and 'action' later became for Karl Marx and Max Weber the central categories of their sociological research. Bubner seeks to clarify how and why it was possible for a sociological theory of action to arise which included in its model of action productive work indiscriminately together with other activities. He also points out that this distinction is closely connected with the differentiation made between *techne* (the purely technical and practical application of knowledge in the context of the means–end relationships of instrumental action, see p. 103ff.) and political action (knowledge is also applied during a dialogue on the subject of the goal of an action – for which means are to be deployed). Aristotle's question was whether or not political action may be subject to the criteria of technical (instrumental) action. In the terminology of modern debate: should sociology be seen as 'social technology' or as a 'theory of communicative competence'? (see Habermas and Luhmann 1979).
2 For Parsons (see Jensen 1980: 57f.) the necessity for action orientation within the framework of a selection grid arises from the fact that the agent is too 'open to the world' and too undetermined by nature to cope without any orientation, i.e. if he had mere stimulus-response and the corresponding mechanism of instinct.

3 Every situation is classified according to the perspective of the purpose of the act. Its representation depends on the one hand on the classification of the goal orientation. On the other, it is assumed that every action situation is selectively relevant for the agent with regard to the goal of the action.

4 Although it occasionally leads to repetitions, this analytical procedure of ethnomethodological action models favoured by Garfinkel (1967), Cicourel (1964) and others is just as acceptable as that of Berger and Luckmann (1966) and Zaner (1961), for the following reasons: first, because Schutz alone consistently uses basic epistemological postulates of the subjective perspective on the specialist theoretical level (see Bernstein 1976: 135); second, because all the sociologists mentioned above refer more or less systematically to the pioneering work of Schutz, so that an analysis of their ideas would be a detour; third, because ethnomethodological concepts can only be fruitful as a starting point for an action theory of social geography if the more general preconditions for their application have been clarified.

5 See Gäfgen 1974: 30ff.

6 See Neumann and Morgenstern 1966.

7 An important sub-group of logical action comprises economic action in the context of the neo-classical theory of economy, or rather its founders Gossen, Menger, Jevons and others. Weber (1968: 63ff.) makes a distinction here between economic action and other forms of logical action in that the former is oriented towards a 'subjectively perceived scarcity' (1968: 65) of means available for action to realize the intention of 'efficiency'. For a perspective of the application in social geography, it is sufficient for the moment to point out that the models of von Thünen, Alfred Weber, Christaller and Lösch take this logical-economic action in principle as a starting point. These models can be seen as further theoretical differentiations of action which take account of spatial distance (i.e. the cost of covering them) and determine the content of these actions within the framework of the corresponding models: the economics of agricultural action in von Thünen's model, industrial action in Weber's model, and service industry action in Christaller and Lösch's model.

8 See Chapter 2 for Popper's distinction between 'formal and empirical rationality'. In the formal aspect, all human actions are rational ('logical' in Pareto's terminology). Every agent works on the basis of his subjectively available general knowledge, and applies it to the specific situation, deducing from it a choice of means for the achieving of a particular goal. Only this aspect is significant for Pareto in the subjective perspective.

9 See Pareto 1917: 127–8 and 1980: 20, 31ff., 113ff.

10 Without going into all the ideal-typical categories of action in detail, suffice it to give the following brief description – see Weber 1968: 24ff. – which is intended to clarify the conditions for the classification of a particular action into one of the types.

11 Parsons places value/norm-oriented action at the centre of his theories; see the next section of this chapter.

12 What Weber terms affectual or traditional action is described by Pareto as non-logical acts.

13 See Pareto 1980: 5ff., 19ff.; 1971: 5ff., 29ff.

14 By 'value system' a large number of rules for evaluation is meant which must be clearly distinguished from Parsons' system of cultural values ('cultural system', see next section of this chapter). Gäfgen (1974: 99f.) sees in it the reason for a clear differentiation between the norm-oriented and the logical model of action. Parsons' argument (1952) that as a rule an agent cannot act logically because he always has to comply with a series of social norms and cultural values, is considered by Gäfgen to be logically contradictory. For 'either the agent has deeply internalized the social norms so that they are part of his [evaluation rules], or their observance is enforced by social sanctions' (1974: 99). Thus in the first case the choice-theory concept of 'evaluation rules' is adequate. In the second, according to Gäfgen, we have merely an expected (negative) consequence of a choice of alternatives which the agent must take into account or evaluate as cost incurred, before he decides upon an alternative. Accordingly, choice theoreticians consider a norm-oriented model of action to be superfluous.

15 See Pareto 1980: 25ff. The distinction between A and C has much in common with Popper's three-world model of knowledge theory. Here A corresponds to world 2, and C to world 3. This suggestion needs further investigation, however.

16 In this context a distinction has to be made between the descriptive norm-oriented and the prescriptive norm-oriented models of action. The writers cited here and the schools of research connected with them can be identified with the descriptive conception. Prescriptive conceptions are those which state what should be, i.e. prescribe ethical standards, and measure action accordingly (see Rawls 1971; Höffe 1981). The action model of the 'philosophy of practice' of Schwemmer (1971, 1979) within the framework of the constructive theory of science of the Erlangen School (Lorenzen 1974) also belongs here.

17 For the German debate, see Mühlmann 1938: 91ff.; Thurnwald 1939: 8ff.; and Luhmann 1962.

18 In the terminology of action theory 'manifest functions' are termed 'intended consequences of actions' and 'latent functions' are 'unintended consequences of actions'.

19 See Dahrendorf 1977.

20 Needs perceived in this way also represent the action-motivating 'strength'.

21 According to Münch (1987: 17ff.), 'interpenetration' is Parsons' central theory. Just as a new quality emerges on a higher level, out of the synthesis of thesis and antithesis (cf. Hegel), the theory of interpenetration encompasses the process whereby opposites are reconciled and the threshold of incompatibility is thrust further into the distance than it was before (see Münch 1987: 27–8). According to Parsons, interpenetration of subsystems results in a qualitatively new

unfolding (of culture, society and personality) by means of a performed act.

22 Quoted from Münch 1987: 17.

23 See Münch 1987: 37ff. In this context Parsons' social theory is often described as a 'concept of institutionalized individualism', where the 'institutionalization of a normative order and the individualization of the personality do not exclude each other but reinforce each other' (Münch 1987: 24), and where the institutionalization is seen as an internalization of norms and values.

24 Quoted from Schutz and Parsons 1978: 29.

25 His definition of 'norm' should also be considered here. It points to the descriptive character of his analysis, and to the difference between his own and prescriptive approaches. 'A norm is a verbal description of the concrete course of action thus regarded as *desirable*, combined with an injunction to make certain future actions conform to this course' (Parsons 1937: 75; quoted from Schutz and Parsons 1978: 13).

26 See Münch 1987: 35. According to Parsons, the processes of institutionalization and internalization mainly take place through the formation of value consensus, so that a mutual orientation of action is intersubjectively possible (Münch 1987: 131–2).

27 In this context, economic action is no longer merely (as for Weber 1968: 63ff.) logical action occurring where there is shortage of means. For Parsons, economic action appears to be an action which is also subject to social norms and cultural values, so that the 'socio-cultural regulation of economic action by value commitments means [. . .] that general value commitments are the means by which normative limits are set for economic action, and that the economic action receives a positive normative orientation' (Münch 1987: 134–5). Calculation of benefit must be linked to (internalized/ institutionalized) norms, so that a voluntaristic social order can exist. The bridge between the socio-cultural and the economic context can be established through practical discourse. 'Should such practical discussions not take place, or remain confined simply to the ideal or the economic plane, the socio-cultural and economic actions will grow apart' (Münch 1987: 135), resulting in a 'pathological'/ 'dysfunctional' (see also Durkheim 1984) social order.

28 A term first used by Freud (1972) and adopted by Parsons. The original German term is *Besetzung*.

29 It should be noted here that each of the four basic functions is a specific expression of one of the aspects of action. A: aspect of the (normative) mobilization of means, which fulfils the adaptation function. G: aspect of goal-attainment, which may be orientated diffusely or specifically towards a clearly defined goal. I: aspect of integration, which involves the interpenetration of the various subsystems. L: aspect of action which involves the retention of latent action patterns.

30 It should be noted that the work of Garfinkel (e.g. 1967), specifically addresses Parsons' abstract interest in action, using the work of Schutz explicitly to develop ethnomethodology's action theory, but

not to the sorts of applications to space that concern me here (see also Heritage 1984; Boden 1990).

31 In answering these questions it is inevitable that we repeat what was already covered in Chapter 3, if Schutz's specialist argumentation is not to be interspersed with too many references to previous sections.

32 According to Schutz (1966: 120), the following example can be used: an engineer has 'knowledge about' of the telephone which he installs. He is an expert on telephones. He knows *how* they function and *why* they function, and on what principles they are constructed. On the basis of this knowledge he is in a position to repair a telephone where necessary.

33 On the other hand, as a non-specialist I can still use the telephone. If it is defective, I apply to the expert. 'What happens when we operate the dial of the telephone is unknown to the non-expert, it is incomprehensible to him and even immaterial; it suffices that the partner to whom he wants to speak answers the telephone' (Schutz 1966: 120).

34 It is not clear here whether 'place' means merely the geographical position, which is interchangeable, or whether it refers to the social position, the same stock of knowledge, etc. As far as the optical dimension is concerned, Schutz seems to be referring merely to the geographical position.

35 Schutz (1970: 13) defines the term 'relevance' as follows: 'The theory concerning the mind's selective activity is simply the title for a set of problems [. . .] – namely, a title for the basic phenomenon we suggest calling "Relevance".' 'Relevance' refers thus in its most general sense to a selection activity of the intellect, or of the consciousness. As we shall see, Schutz distinguishes various forms of selection activity, which are related to different categories of conscious activity: the topical relevance of an object (thematic relevance), the importance of that object for the agent (interpretative relevance), and the incorporation of the object into the purpose of the agent (motivational relevance).

36 See Schutz 1970: 20, 24–5, 37–8, 40–1, 50, 53ff.

37 This example indicates that the following account of the frame of reference of orientation is initially limited to a particular area of action orientation: the area of (constitutive) perception. This limitation has been chosen on the one hand for reasons of simplicity, and on the other because Schutz (1970) bases his analysis on this problem area. In particular, orientation should be considered from the point of view of various dimensions of reality, especially in regard to the social world, communication, and interaction with fellow human beings (see 1970: 167ff.).

38 Two meanings of 'horizon' must be distinguished here (see Schutz 1970: 30–1). First, the external horizon. This comprises everything that appears together with the topic in the immediate field of consciousness. In the example this means all the objects in the room, the spatial and temporal dimension of the action situation. Second, the inner horizon. This comprises everything within the topic in question.

'Once the theme has been constituted, it becomes possible to enter more and more deeply [. . .] into its structure. [. . .] The theme [. . .] is therefore itself an unlimited field for further thematizations' (Schutz 1970: 31). In the example, this refers to anything 'which has to do with the coil as a rope or as a snake'.

39 See Gurwitsch 1971: xxi f. and Lûchinger 1982: 123.
40 For proof of further agreement between knowledge theory and the methodological viewpoint, see Pareto 1971: 5–6; 8ff., 11ff., 18.

6 Geographical space and society

1 See Bartels (1970b, c) and Wirth (1979) for the German and Gregory (1981) and Thrift (1983) for the Anglo-Saxon debate.
2 See Sedlacek 1982: 192–3.
3 See Sorokin 1964: 97ff.
4 See Kutschera 1972: 22ff.
5 The following will serve as an example. I have before me a material artifact which anyone without further ado would describe as a 'table'. The combination of letters 't-a-b-l-e' is generally defined as follows, i.e. determined by the following feature dimensions which an object must fulfil if it is to be described as a 'table' according to the general understanding of the term: upright stand with a horizontal top, not for sitting on. All objects having these features can be assigned to the class of objects defined by their content as 'tables'. We are thus in a position to classify everything logically by means of terms and their definitions. In this example it is noticeable that the particular point in space occupied by this material artifact does not constitute a characteristic feature which has to be fulfilled before the object can be described as a 'table'.
6 See Derrida 1978.
7 See Chapter 2.
8 See Bartels 1974: 13; Sack 1980.
9 See Sack 1980.
10 In principle this is a space-related view of the social frame of reference of action orientation (see Chapter 5) and of the changes wrought by actions in that world. As Parsons and Bales (1953: 85) write, their attention was first drawn by R. Bush to the fact that the AGIL scheme could be seen as a 'four-dimensional space in the mathematical sense of that term'.
11 The four dimensions are seen as comprising the functions of the action orientation of the AGIL scheme 'Adaptation' (A), 'Goal-Attainment and Goal-Selection' (G), 'Integration' (I) and 'Latent Pattern Maintenance' (L). See Parsons and Bales 1953: 88ff.
12 La théorie marxiste des classes [. . .] resultent du fait qu'en réduisant le monde sociale au seul champ économique, elle se condamme à définir la position sociale par référence à la seule position dans les rapports de production économique et elle ignore du même coup les positions occupées dans les différents champs et sous champs. Elle se donne ainsi un monde sociale unidimensionel, simplement organisé

autour de l'opposition entre deux blocs (propriétaires et non propriétaires des moyens de production économique), (Bourdieu 1984a: 9).

13 'Ces classes [. . .] permettent d'*expliquer* et de prévoir les pratiques et les propriétés des choses classées: et entre autre choses les conduites de rassemblement de groupe' (Bourdieu 1984a: 4).

14 'Des classes sur le papier' (Bourdieu 1984a: 4).

15 'Cet espace est aussi réel qu'un espace géographique' (Bourdieu 1984a: 4).

16 'La même chose voudriat pour les rapports entre l'espace géographique et l'espace sociale: ces deux espaces ne coincident jamais complètement; cependant nombre de différence que l'on associe d'ordinaire à l'effet de la distance dans l'espace social, par exemple à l'opposition entre le centre et la périphérie, c'est à dire de la distribution inégale des différentes espèces de capital dans l'espace géographique' (Bourdieu 1984a: 4).

17 'Les déplacements se paient en travail, en efforts et surtout en temps' (Bourdieu 1984a: 4).

18 'Inversement proportionelle à l'éloignement dans cet espace sociale' (Bourdieu 1984a: 4).

19 'Le champ du pouvoir est l'espace des rapports de forces entre des agents ou des institutions ayant en commun de posséder le capital nécessaire pour occuper des positions dominantes dans les différents champs (économique ou culturel notamment)' (Bourdieu 1991a: 5).

20 'Analyser les habitus des occupants de ces positions' (see also Bourdieu 1984a: 11).

21 'Les rapports des autres champs au champ de production sont à la fois des rapports [. . .] de dépendance causale, la forme des determinations causales étant definie par les relations structuales' (Bourdieu, 1984a: 10).

22 'C'est la strucutre du champ politique, c'est-à-dire la relation objective [. . .] des positions [. . .] qui [. . .] détermine les prises de position, c'est-à-dire l'offre des produits politiques' (Bourdieu 1984a: 10).

23 Kolaja (1969: 31) also comments that clearly 'Parsons and Bales have not elaborated the coordination between the action space and the behavioural space'.

24 See Chapter 5.

25 In this context it is worth pointing out the difference between the sphere of investigation of landscape geography and the spatial concept of the subjective perspective. The former starts out from the idea that man carries the totality of his experience of space around with him like a snail in its house, whereas, according to the latter, spatial extension exists independently of the knowing subject: it acquires meaning, however, only through the position and the intentions of the agent. According to the second idea, the agent does not carry 'space' around with him. What actually moves is the point of origin of the coordinates to which the categories of orientation relate. Thus, in Schutz's perspective, 'space' is not reified.

26 For other situations the objective spatial frame of reference is far

more effective. This is the reason for the above-mentioned critique on Bollnow's and Lefebvre's considerations.
27 See Sorokin 1964: 141ff.; Bollnow 1980: 63ff.; Leemann 1976; Dürr 1983.
28 This is exactly the mistake that Zelinsky (1973: 73ff.) is producing in his cultural geography.

7 The space of social theory

1 See Simmel 1903: 43ff.
2 Translated by the author. The original quote is this: 'ces concepts expriment la manière, dont la société se représente les choses'.
3 See Leeman (1976), who confirms Durkheim's theory in his analysis of the connections between world view and spatial concepts in Bali.
4 There is extensive sociological literature on the topic of the social content of material artifacts, association brought about by physical material, although it is hardly taken into account in current discussions, on action theory. Representative authors are: Freyer (1923), Schmalenbach (1927), Gäfgen (1955), Linde (1972) and Lenk and Ropohl (1978). Since their publications largely deal with theories already contained in *The Rules of Sociological Method* (Durkheim 1982), I shall be concentrating here primarily on Durkheim's work. My main intention is to investigate more closely the social structuring of the spatial dimension of the physical world by immobile artifacts, an idea already suggested by Simmel.
5 Revised translation by me. In the French original Durkheim's expression is 'milieu humain' and the English translation is totally misleading by using 'human environment'.
6 Further investigations of this should draw upon Braudel's distinction between 'events' [*conjonctures*] and 'long-duration' [*histoire de la longue durée*]. See Braudel (1972/3: 20–4; 1980: 10–13, 27–34, 74).
7 Of particular note, among others, are the authors Siewert (1972), Atteslander and Hamm (1974), Treinen (1965, 1974), Bourdieu (1974), Greverus (1979) and Hamm (1982). Since the basic ideas stem from Simmel, Pareto and Halbwachs, however, and Pareto's ideas on this topic are dealt with in detail in the following section, I shall concentrate first on the arguments of Simmel and Halbwachs.

8 The space of social action

1 See Pareto 1917: 450.
2 The aggregates of individual parts (persons, objects, artifacts, relationships with people living or dead) often have a proper name, whereby they acquire a degree of consistency so great that they are easily personified. For this to happen it is necessary for aggregates to persist for a long period of time. It is only then that the names of personified abstractions, which do not exist objectively as '*êtres réels*', acquire a high level of content for agents' orientation in their (non-

logical) actions, i.e. *via* the emotional relations which the agent has developed towards them.

3 The original term in German is '*Anlass*' and the standard translation by Roth, Parsons and others is completely misleading. They translated '*Anlass*' as 'stimuli', which I can only interpret as an expression of the power of behavioural thinking in the context of American sociology. Together with the tendency to turn Weber into a functionalist, this is maybe an important obstacle for an adequate Weber interpretation.

4 The standard translation of 'Produktionsrichtung' is again misleading. 'Productive function' is not an adequate term.

5 See von Thünen 1910.

6 See Alfred Weber 1909.

7 See Christaller 1980.

8 See Schilling-Kaletsch 1976: 7ff., 64ff.

9 Parsons defines 'territorial location' as follows: 'Given spatial position of *ego* as *organism* at a given time, the relation to this of the position in space at which *alter ego* is located.'

10 The first context in which territorial location becomes relevant for Parsons (1952: 91) reads like a summary of the 'time-geography' of Hägerstrand, his collaborators and pupils. Since these geographers nowhere cite Parsons as at least a possible source of inspiration, I have quoted the passage in full. My comparison of the time-geography standpoint with the work of Parsons was based on the following publications: Hägerstrand (1970; 1977), Carlstein *et al.* (1978a, b), Pred (1977), Parkes and Thrift (1980). Hanson and Hanson (1980) and Tivers (1977), who claim to have expanded Hägerstrand's time-geography to include the role aspect, seem not to have taken note of Parsons, or at least do not admit to having done so. The parallel nature of the ideas is here considered as a starting point for the integration of time-geography into action-oriented geography.

11 In contrast to Berger and Luckmann (1966), I shall be including Schutz's 'theory of life-forms' in my analysis of this relationship.

12 See Chapter 3.

Bibliography

Adorno, Th.W. (1982) *Against Epistemology: A Metacritique: Studies in Husserl and the Phenomenological Antinomies*, tr. W. Domingo, Oxford: Blackwell.

Adorno, Th.W., Albert, H., Dahrendorf, R., Habermas, J., Pilot, H. and Popper, K.R. (eds) (1976) *The Positivist Dispute in German Sociology*, London: Heinemann.

Agassi, J. (1960) 'Methodological individualism', *The British Journal of Sociology*, 11, 3: 244–70.

Albert, H. (ed.) (1964) *Theorie und Praxis*, Tübingen: J.C.B. Mohr (Paul Siebeck).

Albert, H. (ed.) (1972) *Theorie und Realität*, Tübingen: J.C.B. Mohr (Paul Siebeck).

Alexander, H.G. (ed.) (1956) *The Leibniz-Clarke Correspondence*, Manchester: Manchester University Press.

Althusser, L. and Balibar, E. (1970) *Reading Capital*, London: New Left Books.

Apel, K.-O. (1984) *Understanding and Explanation: A Transcendental-pragmatic Perspective*, tr. G. Warnke, London: MIT Press.

Atteslander, P. and Hamm, B. (eds) (1974) *Materialien zur Siedlungssoziologie*, Köln/Berlin: Kiepenheuer & Witsch.

Bachelard, G. (1986) *La formation de l'esprit scientifique*, 13th edn, Paris: Vrin.

Baeyer, A.v. (1971) 'Einleitung', in A. Schütz *Gesammelte Aufsätze*, vol. 3, Den Haag: Martinus Nijhoff.

Bartels, D. (1968) *Zur wissenschaftstheoretischen Grundlegung einer Geographie des Menschen*, Wiesbaden: Franz Steiner.

Bartels, D. (1970) 'Einleitung', in D. Bartels (ed.) *Wirtschafts- und Sozialgeographie*, Köln/Berlin: Kiepenheuer & Witsch.

Bartels, D. (1973) 'Between theory and metatheory', in R.J. Chorley (ed.) *Directions in Geography*, London: Methuen.

Bartels, D. (1974) 'Schwierigkeiten mit dem Raumbegriff in der Geographie', *Geographica Helvetica*, Beiheft no.2/3: 7–21.

Berger, P.L. and Luckmann, Th. (1966) *The Social Construction of Reality: A Treatise in the Sociology of Knowledge*, London: Allen Lane The Penguin Press.

Bergson, H. (1989) *Zeit und Freiheit*, Frankfurt a.M.: Athenäum.

Bernstein, R.J. (1971) *Praxis and Action*, Philadelphia: University of Pennsylvania Press.

Bernstein, R.J. (1976) *The Restructuring of Social and Political Theory*, Oxford: Basil Blackwell.

Berry, B.J.L. (1971) 'Die wechselseitigen Abhängigkeiten zwischen Bewegungen im Raum und räumlichen Strukturen. Zur Grundlegung einer allgemeinen Feldtheorie', *Geographische Zeitschrift*, 61, 3: 82–100.

Biemel, W. (1972) Reflexionen zur Lebenswelt-Thematik, in W. Biemel, (ed.) *Phänomenologie heute. Festschrift für Ludwig Landgrebe*, Den Haag: Martinus Nijhoff.

Blumer, H. (1969) *Symbolic Interactionism*, Englewood Cliffs: Prentice-Hall.

Boden, D. (1990) 'The world as it happens: ethnomethodology and conversation analysis', in G. Ritzer (ed.) *Frontiers of Social Theory: Toward a New Synthesis*, New York: Columbia University Press.

Bollnow, O.F. (1980) *Mensch und Raum*, 4th edn, Stuttgart/Berlin/Köln/Mainz: Kohlhammer.

Bonss, W. (1982) *Die Einübung des Tatsachenblicks. Zur Struktur und Veränderung empirischer Sozialforschung*, Frankfurt a.M.: Suhrkamp.

Boudon, R. (1981) *The Logic of Social Action: An Introduction to Sociological Analysis*, tr. D. Silverman and G. Silverman, London: Routledge and Kegan Paul.

Bourdieu, P. (1974) 'Der Habitus als Vermittlung zwischen Struktur und Praxis', in P. Bourdieu *Zur Soziologie der symbolischen Formen*, Frankfurt a.M.: Suhrkamp.

Bourdieu, P. (1984a) 'Espace social et genèse des "classes"', *Actes de la recherche en sciences sociales*, 52%w3–12.

Bourdieu, P. (1984b) 'La perception du monde social: une question de mots?', *Actes de la recherche en sciences sociales*, 52–3: 13–14.

Bourdieu, P. (1984c) 'La représentation de la position sociale', *Actes de la recherche en sciences sociales*, 52–3: 14–15.

Bourdieu, P. (1985) 'Sozialer Raum und Klassen', in P. Bourdieu, *Sozialer Raum und 'Klassen'. Leçon sur la leçon. Zwei Vorlesungen*, Frankfurt a.M.: Suhrkamp.

Bourdieu, P. (1990) *The Logic of Practice*, tr. R. Nice, Cambridge: Polity Press.

Bourdieu, P. (1991a) 'Le champs littéraire', *Actes de la recherche en sciences sociales*, 89, 7: 3–46.

Bourdieu, P. (1991b) *Language and Symbolic Power*, ed. and intr. J.B. Thompson, tr. G. Ray and M. Adamson, Cambridge: Polity Press.

Braudel, F. (1972/3) *The Mediterranean and the Mediterranean World in the Age of Philip II*, 2 vols, London: Harper & Row Publishers.

Braudel, F. (1980) *On History*, Chicago/London: University of Chicago Press.

Brauner, H. (1978) *Die Phänomenologie Edmund Husserls und ihre Bedeutung für soziologische Theorien*, Meisenheim a.G.: Hain.

Brennan, T. (ed.) (1989) *Between Feminism and Psychoanalysis*, London: Routledge.

Brodbeck, M. (1968) 'Methodological individualism: Definitions and reductions', in M. Brodbeck (ed.) *Readings in the Philosophy of Social Sciences*, London: Macmillan.

Bubner, R. (1982) *Handlung, Sprache und Vernunft*, Frankfurt a.M.: Suhrkamp.

Bühler, K. (1976) *Die Axiomatik der Sprachwissenschaft*, Frankfurt a.M.: Klostermann.

Bunge, W. (1973) 'Theoretical geography', *Lund Studies in Geography*, no. 1.

Buttimer, A. (1974) 'Values in geography', *Association of American Geographers: Resource Paper*, no. 24.

Buttimer, A. (1976) 'Grasping the dynamism of Lifeworld', *Annals of the Association of American Geographers*, 66, 2: 277–97.

Buttimer, A. (1984) 'Ideal und Wirklichkeit in der Angewandten Geographie', *Münchener Geographische Hefte*, no. 51.

Carlstein, T., Parkes, D. and Thrift, M. (eds) (1978a) *Making Sense of Time*, vol. 1, London: Edward Arnold.

Carlstein, T., Parkes, D. and Thrift, N. (eds) (1978b) *Human Activity and Time Geography*, vol. 2, London: Edward Arnold.

Carnap, R. (1967) *The Logical Structure of the World*, tr. R.A. George, Berkeley: University of California Press.

Carnap, R. (1978) 'Der Raum. Ein Beitrag zur Wissenschaftslehre', *Kantstudien*, no. 56.

Christaller, W. (1980) *Die zentralen Orte in Süddeutschland*, Darmstadt: Wissenschaftliche Buchgesellschaft.

Cicourel, A. (1964) *Method and Measurement in Sociology*, New York: Free Press.

Claesges, U. (1964) *Edmund Husserls Theorie der Raumkonstitution*, Den Haag: Martinus Nijhoff.

Cohen, I. J. (1989) *Structuration Theory: Anthony Giddens and the Constitution of Social Life*, London: Macmillan.

Dahrendorf, R. (1977) *Homo Sociologicus*, 15th edn, Opladen: Westdeutscher Verlag.

Davidson, D. (1980) *Essays on Actions and Events*, Oxford: Claredon Press.

Derrida, J. (1967) *De la grammatologie*, Paris: Editions Seuil.

Derrida, J. (1978) *Writing and Difference*, Chicago: University of Chicago Press.

Descartes, R. (1959) *Principles of Philosophy*, Edinburgh/London: Thomas Nelson and Sons.

Diemer, A. (1956) *Edmund Husserl. Versuch einer systematischen Darstellung seiner Phänomenologie*, Meisenheim a.G.: Hain.

Donagan, A. (1975) 'Die Popper-Hempel-Theorie der historischen Erklärung', in B. Giesen and M. Schmid (eds) *Theorie, Handeln und Geschichte*, Hamburg: Hoffmann & Campe.

Durkheim, E. (1963) *Primitive Classification*, London: Cohen and West.

Durkheim, E. (1971) *The Elementary Form of Religious Life*, London: Allen and Unwin.

Durkheim, E. (1982) *The Rules of Sociological Method and Selected Texts on*

Sociology and its Method, ed. and intr. St Lukes, tr. W.D. Halls, London: Macmillan.

Durkheim, E. (1984) *The Division of Labour in Society*, tr. W.D. Halls, Basingstoke: Macmillan.

Durkheim, E. (1985) *Les formes élémentaires de la vie religieuse*, 6th edn, Paris: Presse Universitaire de France.

Durkheim, E. and Mauss, M. (1901) 'De quelques formes primitives de classification', *L'Année Sociologique*, 6: 1–72.

Dürr, H.P. (1983) *Traumzeit*, Frankfurt a.M.: Syndikat.

Einstein, A. (1954) 'Foreword', in M. Jammer *Concepts of Space*, Cambridge, Mass.: Harvard University Press.

Elster, J. (1982) 'Marxism, Functionalism, and Game Theory: The case for methodological individualism', *Theory and Society*, 11, 4: 453–82.

Esser, H., Klenovits, K. and Zehnpfenning, H. (1977) *Wissenschaftstheorie*, vol. 2, Stuttgart: Teubner.

Foucault, M. (1976) 'Question à Michel Foucault sur la géographie', *Hérodote*, no. 1.

Foucault, M. (1979) *Discipline and Punish*, Harmondsworth: Penguin.

Foucault, M. (1980) *Power/Knowledge: Selected Interviews and other Writings 1972–1977*, Brighton: Harvester.

Foucault, M. (1983) 'How is power exercised?', in H.L. Dreyfus and P. Rabinow (eds) *Michel Foucault. Beyond Structuralism and Hermeneutics*, 2nd edn, Chicago: The University Press of Chicago.

Frege, G. (1892) 'Über Sinn und Bedeutung', *Zeitschrift für Philosophie und philosophische Kritik*, 100, 1: 25–50.

Freud, S. (1972) 'Das Ich und das Es', in S. Freud Gesammelte Werke, Bd. 13, 7. Auflage, Frankfurt a.M.: S. Fischer.

Freyer, H. (1923) *Theorie des objektiven Geistes. Eine Einleitung in die Kulturs-philosophie*, Leipzig/Berlin: Teubner.

Gäfgen, G. (1955) 'Soziologie der Technik', *Kölner Zeitschrift für Soziologie und Sozialpsychologie*, 7, 4: 580–600.

Gäfgen, G. (1974) *Theorie der wirtschaftlichen Entscheidung. Untersuchungen zur Logik und Bedeutung des rationalen Handelns*, Tübingen: J.C.B. Mohr (Paul Siebeck).

Gäfgen, G. (1980) 'Formale Theorie des strategischen Handelns', in H. Lenk (ed.) *Handlungstheorien-interdisziplinär*, vol. 1, München: Wilhelm Fink.

Galtung, J. (1977) 'Essays in methodology', in *Methodology and Ideology*, vol. 1, Copenhagen: Ejlers.

Garfinkel, H. (1967) *Studies in Ethnomethodology*, Englewoods Cliffs: Prentice-Hall.

Geertz, C. (1983) *Local Knowledge: Further Essays in Interpretative Anthro-pology*, New York: Basic Books.

Giddens, A. (1979) *Central Problems in Social Theory: Action, Structure and Contradiction in Social Analysis*, London: Macmillan.

Giddens, A. (1981) *A Contemporary Critique of Historical Materialism: Power, Property and the State*, London: Macmillan.

Giddens, A. (1984) *The Constitution of Society*, Cambridge: Polity Press.

Giddens, A. (1985) 'Time, space and regionalisation', in D. Gregory and J. Urry (eds) *Social Relations and Spatial Structures*, London: Macmillan.

Giddens, A. (1990) *The Consequences of Modernity*, Standford: Standford University Press.

Giddens, A. (1991) *Modernity and Self-Identity: Self and Society in the Late Modern Age*, Cambridge: Polity Press.

Girndt, H. (1967) *Das soziale Handeln als Grundkategorie erfahrungswissenschaftlicher Soziologie*, Tübingen: J.C.B. Mohr (Paul Siebeck).

Gold, J.R. (1980) *An Introduction to Behavioural Geography*, Oxford: Oxford University Press.

Gouldner, A.W. (1973) 'Reziprozität und Autonomie in der funktionalen Theorie', in H. Hartmann (ed.) *Moderne amerikanische Soziologie*, Stuttgart: Enke.

Grathoff, R. (1978a) 'Alfred Schütz', in D. Kaesler (ed.) *Klassiker des soziologischen Denkens*, vol. 2, München: C.H. Beck.

Grathoff, R. (1978b) 'Alltag und Lebenswelt als Gegenstand der phänomenologischen Sozialtheorie', in K. Hammerich *et al.* (eds): *Materialien zur Soziologie des Alltags*, Opladen: Westdeutscher Verlag.

Gregory, D. (1978) *Ideology, Science and Human Geography*, London: Hutchinson.

Gregory, D. (1981) 'Human agency and human geography', *Transactions of the Institute of British Geographers: New Series* 6: 1–18.

Gregory, D. and Urry, J. (eds) (1985) *Social Relations and Spatial Structures*, London: Macmillan.

Greverus, I.M. (1979) *Auf der Suche nach Heimat*, München: C.H. Beck.

Gurwitsch, A. (1971) 'Einführung', in A. Schütz *Gesammelte Aufsätze*, vol. 1, Den Haag: Martinus Nijhoff.

Gurwitsch, A. (1974) *Phenomenology and the Theory of Science*, Northwestern University Press.

Habermas, J. (1974) *Theory and Practice*, tr. J. Viertel, London: Heinemann.

Habermas, J. (1984) *The Theory of Communicative Action*, tr. T. McCarthy, vol. 1, Cambridge: Polity Press.

Habermas, J. (1987) *The Theory of Communicative Action*, tr. T. McCarthy, vol. 2, Cambridge: Polity Press.

Habermas, J. (1988) *The Logic of the Social Sciences*, tr. S.W. Nicholsen and J.A. Stark, Cambridge: Polity Press.

Habermas, J. and Luhmann, N. (1979) *Theorie der Gesellschaft oder Sozialtechnologie – Was leistet die Systemforschung?*, Frankfurt a.M.: Suhrkamp.

Hägerstrand, T. (1970) 'What about people in regional science?', *Papers of the Regional Science Association*, 24: 7–21.

Hägerstrand, T. (1977) 'The time impact of social organization and environment upon the time-use of individuals and households', in A. Kulinski (ed.) *Social Issues in Regional Policy and Regional Planning*, The Hague: Mouton.

Hägerstrand, T. (1984) 'Time-geography: Focus on the corporeality of man, society and environment', *Papers of the Regional Science Association*, 31, 193–216.

Halbwachs, M. (1970) *Morphologie sociale*, 2nd edn, Paris: Armand Colin.
Halbwachs, M. (1980) *The Collective Memory*, tr. F.J. Ditter Jr and V.Y. Ditter, London: Harper and Row.
Hamm, B. (1982) *Einführung in die Siedlungssoziologie*, München: C.H. Beck.
Hanson, S. and Hanson, P. (1980) 'Gender and urban activity patterns in Uppsala, Sweden', *Geographical Review*, 70, 3: 291–9.
Hard, G. (1970) 'Die "Landschaft" der Sprache und die "Landschaft" der Geographen', *Colloquium Geographicum*, no. 11.
Hard, G. (1973) *Die Geographie. Eine wissenschaftstheoretische Einführung*, Berlin: De Gruyter.
Hartke, W. (1956) 'Die "Sozialbrache" als Phänomen der geographischen Differenzierung der Landschaft', *Erdkunde*, 10, 4: 257–69.
Hartke, W. (1963) 'Der Weg zur Sozialgeographie', *Mitteilungen der Österreichischen Geographischen Gesellschaft*, 105: 5–17.
Harvey, D. (1973) *Social Justice and the City*, London: Edward Arnold.
Harvey, D. (1985) *The Urbanization of Capital: Studies in the History of Capitalist Urbanization*, Oxford: Basil Blackwell.
Harvey, D. (1989) *The Condition of Postmodernity. An Enquiry into the Origins of Cultural Change*, Oxford: Basil Blackwell.
Hawking, St W. (1988) *A Brief History of Time: From the Big Bang to Black Holes*, London: Bantam Press.
Heidegger, M. (1983) *Die Kunst und der Raum*, 2nd edn, St Gallen: Erker.
Heidegger, M. (1986) *Sein und Zeit*, 16th edn, Tübingen: Max Niemeyer.
Held, K. (1981) 'Edmund Husserl', in O. Höffe (ed.) *Klassiker der Philosophie*, vol. 2, München: C.H. Beck.
Hempel, C.G. (1965) *Aspects of Scientific Explanation*, New York: Free Press.
Hempel, C.G. and Oppenheim, P. (1965) 'Studies in the logic of explanation', in C.G. Hempel *Aspects of Scientific Explanation*, New York: Free Press.
Heritage, J.C. (1984) *Garfinkel and Ethnomethodology*, Cambridge: Polity Press.
Hettner, A. (1927) *Die Geographie. Ihre Geschichte, ihr Wesen, ihre Methoden*, Breslau: Ferdinand Hirt.
Höffe, O. (1980) 'Entscheidungstheoretische Denkfiguren und die Begründung von Recht', in W. Hassemer *et al.* (eds) *Argumentation und Recht*, Wiesbaden: Franz Steiner.
Höffe, O. (1981) *Sittlich-politische Diskurse*, Frankfurt a.M.: Suhrkamp.
Hume, D. (1978) *A Treatise of Human Nature*, ed. with an analytical index by L.A. Selby-Bigge, 2nd edn, Oxford: Clarendon Press.
Husserl, E. (1965) *Philosophy as Rigorous Science. Phenomenology and the Crisis of Philosophy*, tr. Q. Lauer, New York: Harper Torchbooks.
Husserl, E. (1969) *Formal and Transcendental Logic*, tr. D. Cairns, The Hague: Martinus Nijhoff.
Husserl, E. (1970a) *The Crisis of European Sciences and Transcendental Phenomenology*, tr. and intr. C. Carr, Evaston, Ill.: Northwestern University Press.
Husserl, E. (1970b) *Logical Investigations*, tr. J.N. Findlay, London: International Library of Philosophy and Scientific Method.

Husserl, E. (1972) *Erfahrung und Urteil*, 4th edn, Hamburg: Felix Meiner.

Husserl, E. (1976) *Logische Untersuchungen: Über intentionale Erlebnisse und ihre 'Inhalte'*, Hamburg: Felix Meiner.

Husserl, E. (1977) *Phenomenological Psychology: Lectures Summer Semester 1925*, tr. J. Scanlon, The Hague: Martinus Nijhoff.

Husserl, E. (1982) *Ideas Pertaining to a Pure Phenomenology and to a Phenomenological Philosophy*, vol. 1, tr. F. Kersten, The Hague: Martinus Nijhoff.

Jackson, P. (1981) 'Phenomenology and social geography', *Area*, 13: 299–305.

Jackson, P. and Smith, S.J. (1984) *Exploring Social Geography*, London: Allen and Unwin.

James, S. (1984) *The Content of Social Expanation*, Cambridge: Cambridge University Press.

Jammer, M. (1954) *Concepts of Space: the History of Theories of Space in Physics*, Cambridge, Mass.: Harvard University Press.

Jarvie, I.C. (1972) *Concepts and Society*, London/Boston: Routledge and Kegan Paul.

Jensen, S. (1980) *Talcott Parsons*, Stuttgart: Teubner.

Johnston, R.J. (1983) *Philosophy and Human Geography: An Introduction to Contemporary Approaches*, London: Edward Arnold.

Kant, I. (1802) *Physische Geographie*, D.F.Th. Rink (ed.), Königsberg: Gobbels und Unzer.

Kant, I. (1905) 'Neuer Lehrbegriff von Bewegung und Ruhe (1764)', in I. Kant *Gesammelte Schriften*, ed. Königlich Preussische Akademie der Wissenschaften, vol. 2, Berlin: Georg Reimer.

Kant, I. (1956) *Kritik der reinen Vernunft*, Hamburg: Felix Meiner.

Kant, I. (1969) *Critique of Pure Reason*, tr. N.K. Smith, New York: St Martin.

Koertge, N. (1974) 'On Popper's philosophy of the social science', in K. Schaffner and R.S. Cohen (eds) *Proceedings of the 1972 Biennial Meeting of the Philosophy of Science Association*, Dordrecht: Reidel.

Koertge, N. (1975) 'Popper's metaphysical research program for the human sciences', *Inquiry*, 18: 437–62.

Koertge, N. (1979) 'The methodological status of Popper's rationality principle', *Theory and Decision*, 10: 83–95.

Kolaja, J. (1969) *Social System and Time and Space: An Introduction to the Theory of Recurrent Behaviour*, Pittsburgh: Greenwood.

Konau, E. (1977) 'Raum und soziales Handeln. Studien zu einer vernachlässigten Dimension soziologischer Theoriebildung', *Göttinger Abhandlungen zur Soziologie*, vol. 25, Stuttgart: Enke.

Kuhn, Th.S. (1970) *The Structure of Scientific Revolutions*, 2nd edn, Chicago: University of Chicago Press.

Kutschera, F. (1972) *Wissenschaftstheorie*, 2 vols, München: UTB Wilhelm Fink.

Küng, G. (1982) *Einführung in die Phänomenologie*, unpublished manuscript, Fribourg.

Landgrebe, L. (1963) *Der Weg der Phänomenologie. Das Problem der ursprünglichen Erfahrung*, Gütersloh: Gütersloher Verlagshaus G. Mohn.

Leemann, A. (1976) 'Auswirkungen des balinesischen Weltbildes auf verschiedene Aspekte der Kulturlandschaft und auf die Wertung des Jahresablaufes', *Ethnologische Zeitschrift Zurich*, 2: 27–67.

Lefebvre, H. (1981) *La production de l'espace*, 2nd edn, Paris: Editions Anthropos.

Leibniz, G.W. (1904) *Hauptschriften zur Grundlegung der Philosophie*, ed. E. Cassirer and tr. A. Buchenau, 2 vols, Leipzig: F. Meiner.

Lenk, H. and Ropohl, G. (1978) 'Technik im Alltag', in K. Hammerich and M. Klein (eds) *Materialien zur Soziologie des Alltags*, Opladen: Westdeutscher Verlag.

Ley, D. (1977) 'Social geography and the taken-for-granted-world', *Transactions of the Institute of British Geographers: New Series*, 2: 498–512.

Ley, D. (1978) 'Social geography and social action', in D. Ley and M. Samuels (eds) *Humanistic Geography*, London: Croom Helm.

Lichtman, R. (1965) 'Karl Popper's defense of the autonomy of sociology', *Social Research*, 32: 1–25.

Linde, H. (1972) *Sachdominanz in Sozialstrukturen*, Tübingen: J.C.B. Mohr (Paul Siebeck).

Lorenzen, P. (1974) *Konstruktive Wissenschaftstheorie*, Frankfurt a.M.: Suhrkamp.

Lüchinger, N. (1982) *Phänomenologie der Lebenswelt und Soziologie des Alltags*, unpublished thesis, Fribourg.

Luckmann, Th. (1978) *Poetik und Hermeneutik. Zum hermeneutischen Problem der Handlungswissenschaften*, unpublished manuscript, Konstanz.

Luckmann, Th. (1982) 'Einleitung', in A. Schutz *Das Problem der Relevanz*, Frankfurt a.M.: Suhrkamp.

Luckmann, Th. (1983) *Life-World and Social Realities*, London: Heinemann Educational.

Luhmann, N. (1962) 'Funktion und Kausalität', *Kölner Zeitschrift für Soziologie und Sozialpsychologie*, 14, 4: 617–644.

Lukes, S. (1977) 'Methodological individualism reconsidered', in *Essays in Social Theory*, London: Macmillan.

Lukes, S. (1979) *Individualism*, Oxford: Basil Blackwell.

Lyotard, J.F. (1986) *Das postmoderne Wissen. Ein Bericht*, Wien: Edition Passagen.

Malinowski, B. (1990) *A Scientific Theory of Culture and other Essays*, 9th edn, Chapel Hill, N.C.: University of North Carolina Press.

Marx, K. (1960) *Max Engels Werke*, vol. 8, Berlin: Dietz.

Marx, K. (1962) *Das Kapital. Kritik der politischen Ökonomie*, vol. 1, Berlin: Dietz.

Maus, H. (1967) 'Geleitwort', in M. Halbwachs *Das kollektive Gedächtnis*, Stuttgart: Enke.

Mead, G.H. (1967) *Mind, Self, and Society*, Chicago: University of Chicago Press.

Mellor, D.H. (1981) *Real Time*, Cambridge: Cambridge University Press.

Merton, R. (1957) 'Manifest and latent functions', in R. Merton *Social Theory and Social Structure*, Glencoe, Ill.: Free Press.

Mill, J.St (1963) *Collected Works*, London: Routledge and Kegan Paul.

Mill, J.St (1973) *A System of Logic*, ed. J.M. Robson and R.F. McRae, vol. VI, London: Routledge and Kegan Paul.

Mischel, Th. (1981) *Psychologische Erklärungen. Gesammelte Aufsätze*, Frankfurt a.m.: Suhrkamp.

Mitscherlich, A. (1970) *Die Unwirtlichkeit unserer Städte. Anstiftung zum Unfrieden*, Frankfurt a.m.: Suhrkamp.

Mongardini, C. (1975) 'Paretos Soziologie um die Jahrhundertwende', in V. Pareto, *Ausgewählte Schriften*, Frankfurt a.M./Berlin/Wien: Ullstein.

Moya, C. J. (1990) *The Philosophy of Action: An Introduction*, Cambridge: Polity Press.

Mühlmann, W. (1938) *Methodik der Völkerkunde*, Stuttgart: Enke.

Münch, R. (1987) *Theory of Action: Towards a new Synthesis going beyond Parsons*, London: Routledge and Kegan Paul.

Nerlich, G. (1976) *The Shape of Space*, Cambridge: Cambridge University Press.

Neumann J.v. and Morgenstern, O. (1966) *Theory of Games and Economic Behaviour*, 3rd edn, Princeton N.J.: Princeton University Press.

Newton, I. (1934) *Mathematical Principles of Natural Philosophy*, Berkeley: University of California Press.

Newton, I. (1952) *Opticks or a Treatise of the Reflections, Refractions, Inflections and Colours of Light*, new edn, New York: Dover Publications.

Opp, K.O. (1979) *Methodologie der Sozialwissenschaften*, Reinbek bei Hamburg: Rowohlt.

Övermann, U. *et al.* (1979) 'Die Methodologie der objektiven Hermeneutik und ihre allgemeine forschungslogische Bedeutung in den Sozialwissenschaften', in H.G. Söffner (ed.) *Interpretative Verfahren in den Sozialwissenschaften*, Stuttgart: Metzlersche Verlagsbuchhandlung.

Pareto, V. (1909) *Manuel d'économie politique*, Genève: Droz.

Pareto, V. (1910) 'Le azioni non logiche' *Rivista italiana di sociologia*.

Pareto, V. (1917) *Traité de Sociologie Générale*, Paris/Genève: Droz.

Pareto, V. (1971) *Manual of Political Economy*, tr. A.S. Schwier and A.N. Page, Basingstoke: Macmillan.

Pareto, V. (1975) 'Eine Anwendungsform soziologischer Theorien', in V. Pareto, *Ausgewählte Schriften*, Frankfurt a.M./Berlin/Wien: Ullstein.

Pareto, V. (1980) *Compendium of General Sociology*, ed. E. Abbot, Minneapolis: University of Minnesota Press.

Parkes, D.N. and Thrift, N.J. (1980) *Times, Spaces and Places: a chronographic perspective*, Chichester/New York/Brisbane/Toronto: John Wiley and Sons.

Parsons, T. (1937) *The Structure of Social Action*, New York: McGraw-Hill.

Parsons, T. (1952) *The Social System*, London: Free Press.

Parsons, T. (1960) 'The principal structures of community', in *Structure and Process in Modern Societies*, Glencoe, Ill.: Free Press.

Parsons, T. (1964) 'Die jüngsten Entwicklungen in der strukturell-funktionalen Theorie', *Kölner Zeitschrift für Soziologie und Sozialpsychologie*, 16, 1: 30–49.

Parsons, T. (1967) 'Pattern variables revisited: A response to Robert Dubin', in T. Parsons, *Sociological Theory and Modern Society*, New York: Free Press.

Parsons, T. (1978) *Action Theory and the Human Condition*, New York: Free Press.

Parsons, T. and Bales, R.F. (1953) 'The dimension of action space', in T. Parsons, R.F. Bales and E.A. Shils *Working Papers in the Theory of Action*, Glencoe, Ill.: Free Press.

Parsons, T., Bales, F.B. and Shils, E.A. (1953) *Working Papers in the Theory of Action*, Glencoe, Ill.: Free Press.

Parsons, T. and Shils, E.A. (eds) (1951) *Toward a General Theory of Action*, Cambridge, Mass.: Harvard University Press.

Peursen, C.A.v. (1969) *Phänomenologie und analytische Philosophie*, Stuttgart: Kohlhammer.

Pickles, J. (1985) *Phenomenology, Science and Geography: Spatiality and the Human Sciences*, Cambridge: Cambridge University Press.

Pivčević, E. (1970) *Husserl and Phenomenology*, London: Hutchinson.

Piveteau, J.L. (1983) 'Un certain chant du monde', *Cahiers de l'Institut de Géographie de Fribourg*, 1: 5–9.

Popper, K.R. (1933) 'Ein Kriterium des empirischen Charakters theoretischer Systeme', *Erkenntnis*, 3, 1: 426–7.

Popper, K.R. (1960) *The Poverty of Historicism*, 2nd edn, London: Routledge and Kegan Paul.

Popper, K.R. (1967) 'La rationalité et le status du principe de rationalité', in E.M. Classen (ed.) *Les fondements philosophiques des systèmes économiques*, Paris: Payot.

Popper, K.R. (1968) *The Logic of Scientific Discovery*, 2nd edn, London: Hutchinson.

Popper, K.R. (1969) *The Open Society and its Enemies*, vol. 2, 5th edn, London: Routledge and Kegan Paul.

Popper, K.R. (1976) 'The logic of social sciences', in Th.W. Adorno *et al.* (eds) *The Positivist Dispute in German Sociology*, London: Heinemann.

Popper, K.R. (1979) *Objective Knowledge. An Evolutionary Approach*, rev. edn, Oxford: Claredon Press.

Popper, K.R. (1982) *Die Logik der Forschung*, 7th edn, Tübingen: J.C.B. Mohr (Paul Siebeck).

Popper, K.R. (1986) *Unended Quest: An Intellectual Autobiography*, London: Flamingo, Fontana Paperbacks.

Pred, A. (1977) 'The choreography of existence: comments on Hägerstrands time-geography and its usefulness', *Economic Geography*, 53, 2: 207–21.

Prewo, R. (1979) *Max Webers Wissenschaftsprogramm. Versuch einer methodischen Neuerschliessung*, Frankfurt a.M.: Suhrkamp.

Prim, R. and Tilmann, H. (1979) *Grundlagen einer kritisch-rationalen Sozialwissenschaft*, Heidelberg: UTB Quelle and Meyer.

Radcliffe-Brown, A.R. (1952) *Structure and Function in Primitive Society*, London: Free Press.

Raffestin, C. (1986) 'Territorialité: Concept ou paradigme de la géographie sociale', *Geographica Helvetica*, 2: 91–6.

Rawls, J. (1971) *A Theory of Justice*, Cambridge, Mass.: Harvard University Press.

Rombach, H. (ed.) (1974) *Wissenschaftstheorie*, Freiburg/Basel/Wien: Herder.

Sack, R.D. (1980) *Conceptions of Space in Social Thought. A Geographic Perspective*, London: Macmillan.

Saunders, P. (1985) 'Space, the city and urban sociology', in D. Gregory and J. Urry (eds) *Social Relations and Spatial Structures*, London: Macmillan.

Saussure, F.d. (1960) *Course in General Linguistics*, London: Peter Owen.

Schilling-Kaletsch, I. (1976) 'Wachstumspole und Wachstumszentren. Untersuchungen zu einer Theorie sektoral und regional polarisierter Entwicklung', in *Arbeitsberichte und Ergebnisse zur Wirtschafts- und Sozialgeographischen Regionalforschung*, vol. 1, Universität Hamburg.

Schmalenbach, H. (1927) 'Soziologie der Sachverhältnisse', *Jahrbuch für Soziologie*, 3: 38–45.

Schmid, M. (1979) 'Rationalitätsprinzip und Handlungserklärung', in H. Lenk (ed.) *Handlungstheorien-interdisziplinär*, vol. 2, München: Wilhelm Fink.

Schutz, A. (1962) *Collected Papers*, vol. I, ed. and intr. M. Natason, The Hague: Martinus Nijhoff.

Schutz, A. (1964) *Collected Papers*, vol. II, ed. A. Brodersen, The Hague: Martinus Nijhoff.

Schutz, A. (1966) *Collected Papers*, vol. III ed. I. Schutz, The Hague: Martinus Nijhoff.

Schutz, A. (1970) *Reflections on the Problem of Relevance*, New Haven: Yale University Press.

Schutz, A. (1972) *The Phenomenology of the Social World*, tr. G. Walsh and F. Lehnert, London: Heinemann.

Schutz, A. (1974) *Der sinnhafte Aufbau der sozialen Welt. Eine Einführung in die verstehende Soziologie*, Frankfurt a.M.: Suhrkamp.

Schutz, A. (1982) *Life Forms and Meaning Structures*, tr., intr. and annoted by H.R. Wagner, London/Boston/Melbourne/Henley: Routledge and Kegan Paul.

Schutz, A. and Luckmann, Th. (1974) *Structures of the Life-World*, tr. R.M. Zaner and H.T. Engelhardt Jr, London: Heinemann.

Schutz, A. and Parsons, T. (1978) *The Theory of Social Action: The Correspondence of Alfred Schutz and Talcott Parsons*, ed. R. Grathoff, Bloomington and London: Indiana University Press.

Schwemmer, O. (1971) *Philosophie der Praxis*, Frankfurt a.M.: Suhrkamp.

Schwemmer, O. (1979) 'Verstehen als Methode. Vorüberlegungen zu einer Theorie der Handlungsdeutung', in J. Mittelstrass (ed.) *Methodenprobleme der Wissenschaften vom gesellschaftlichen Handeln*, Frankfurt a.M.: Suhrkamp.

Sedlacek, P. (1982) 'Kulturgeographie als normative Handlungswissenschaft', in P. Sedlacek (ed.) *Kultur-/Sozialgeographie*, Paderborn: UTB Ferdinand Schöningh.

Seiffert, H. (1972) *Einführung in die Wissenschaftstheorie*, vol. 2, München: C.H. Beck.

Siewert, H.-J. (1972) 'Bestimmt die bebaute Umwelt das menschliche Verhalten? Der Raum als Gegenstand der Sozialwissenschaften', *Der Bürger im Staat*, 24, 2: 144–8.

Simmel, G. (1903) 'Soziologie des Raumes', *Jahrbuch für Gesetzgebung, Verwaltung und Volkswirtschaft im Deutschen Reich*, 1, 1: 27–71.

Simmel, G. (1989) *Soziologie. Untersuchungen über die Formen der Vergesell-schaftung*, ed. O. Rammstedt, vol. 11, Berlin: Suhrkamp.

Sklar, L. (1974) *Space, Time and Space-Time*, Berkeley: University of California Press.

Smart, J.J.C. (1964) *Problems of Space and Time*, London: Macmillan.

Soja, E.W. (1989) *Postmodern Geographies. The Reassertion of Space in the Critical Social Theory*, London/New York: Verso.

Sorokin, P.A. (1964) *Sociocultural Causality, Space, Time*, New York: Russel and Roussel.

Spencer, H. (1966/7) *The Works of Herbert Spencer*, vol. 2, Osnabrück: O. Zeller.

Srubar, I. (1979) 'Die Theorie der Typenbildung bei Alfred Schütz', in W.M. Sprondel and R. Grathoff (eds) *Alfred Schütz und die Idee des Alltags in den Sozialwissenschaften*, Stuttgart: Enke.

Srubar, I. (1981) 'Schütz' Bergson-Rezeption', in A. Schütz *Theorie der Lebensformen*, Frankfurt a.M.: Suhrkamp.

Srubar, I. (1988) *Die Genese der pragmatischen Lebenswelttheorie von Alfred Schütz und ihr anthropologischer Hintergrund*, Frankfurt a.M.: Suhrkamp.

Stegmüller, W. (1975) *Hauptstörungen der Gegenwartsphilosophie*, Stuttgart: Kröner.

Szilasi, W. (1959) *Einführung in die Phänomenologie Edmund Husserls*, Tübingen: M. Niemeyer.

Tarski, A. (1969) *Logic, Semantics, Metamathematics: Papers from 1923 to 1938*, tr. J.H. Woodger, Oxford: Claredon Press.

Thomas, W.I. (1967) 'The definition of the situation', in J.G. Manis and B.N. Meltzer (eds) *Symbolic Interaction*, Boston: Allyn and Bacon.

Thrift, N. (1983) 'On the determination of social action in space and time', *Society and Space*, 1: 23–57.

Thrift, N. (1990) 'For a new regional geography 1', *Progress in Human Geography*, 2: 272–9.

Thünen, J.G.H.v. (1910) *Der isolierte Staat in Beziehung auf Landwirtschaft und Nationalökonomie*, Jena: Fischer.

Thurnwald, R. (1939) *Lehrbuch der Völkerkunde*, Stuttgart: Enke.

Tivers, J. (1977) 'Constraints on spatial activity patterns. Women with young children', *Occasional Paper*, no. 6, Department of Geography, Kings College, University of London.

Touraine, A. (1965) *Sociologie de l'action*, Paris: Editions Seuil.

Treinen, H. (1965) 'Symbolische Ortsbezogenheit', *Kölner Zeitschrift für Soziologie und Sozialpsychologie*, 17, 1: 5–73.

Treinen, H. (1974) 'Symbolische Ortsbezogenheit', in P. Atteslander and B. Hamm (eds) *Materialien zur Siedlungssoziologie* Köln/Berlin: Kiepen-heuer & Witsch.

Walmsley, D.J. and Lewis, G.J. (1984) *Human Geography: Behavioral Approaches*, London/New York: Longman.

Watkins, J.W.N. (1972) 'Idealtypen und historische Erklärung', in H. Albert (ed.) *Theorie und Realität*, 2nd edn, Tübingen: J.C.B. Mohr (Paul Siebeck).

Watson, J.B. (1913) 'Psychology as the Behaviorist views it', *Psychological Review*, XX, 1: 158–77.

Watson, J.B. (1970) *Behaviourism*, New York: Norton.

Weber, A. (1909) *Über den Standort der Industrien*, Tübingen: J.C.B. Mohr (Paul Siebeck).

Weber, M. (1913) 'Über einige Kategorien verstehender Soziologie', *Logos*, IV: 253–94.

Weber, M. (1949) *The Methodology of Social Sciences*, tr. and ed. E.A. Shils and H.A. Finck, New York: Free Press.

Weber, M. (1951) *Gesammelte Aufsätze zur Wissenschaftslehre*, Tübingen: J.C.B. Mohr (Paul Siebeck).

Weber, M. (1968) *Economy and Society: An Outline of Interpretive Sociology*, 3 vols (ed. G. Roth and C. Wittich, tr. E. Fischoff *et al.*), New York: Bedminster Press.

Weber, M. (1980) *Wirtschaft und Gesellschaft*, 5th edn, Tübingen: J.C.B. Mohr (Paul Siebeck).

Werlen, B. (1980) *Funktionalismus in Sozialwissenschaft und Geographie*, unpublished thesis, Fribourg.

Werlen, B. (1984) 'Grundkategorien funktionalen Denkens in Sozialwissenschaft und Sozialgeographie', *Cahiers de l'Institut de Géographie de Fribourg*, 1: 1–30.

Werlen, B. (1987) 'Zwischen Metatheorie, Fachtheorie und Alltagswelt', in G. Bahrenberg *et al.* (eds) *Geographie des Menschen – Dietrich Bartels zum Gedenken*, Bremer Beiträge zur Geographie und Raumplanung, no. 11: 11–25.

Werlen, B. (1988a) *Gesellschaft, Handlung und Raum. Grundlagen handlungstheoretischer Sozialgeographie*, 2nd edn, Stuttgart: Franz Steiner.

Werlen, B. (1988b) 'Von der Raum- zur Situationswissenschaft', *Geographische Zeitschrift*, 76, 4: 193–208.

Werlen, B. (1989) 'Kulturelle Identität zwischen Individualismus und Holismus', in K.L. Sosoe (ed.) *Identität: Evolution oder Differenz?*, Fribourg: Editions Universitaire Fribourg Suisse.

Werlen, B. (forthcoming) 'On regional and cultural identity: Outline of a regional culture analysis', in D. Steiner and M. Nauser (eds) *Person, Society, Environment*, London: Routledge.

Wirth, E. (1979) *Theoretische Geographie*, Stuttgart: Teubner.

Wright, G.H.v. (1971) *Explanation and Understanding*, London: Routledge and Kegan Paul.

Zaner, R.M. (1961) 'Theory of Intersubjectivity: Alfred Schutz', *Social Research*, 28: 71–91.

Zelinsky, W. (1973) *The Cultural Geography of the United States*, Englewood Cliffs, N.J.: Prentice-Hall.

Name index

Subject index

abstraction: and social relationships 171

action 2, 3, 5, 6–7, 8, 9–20, 24, 25, 28, 30, 41, 43, 44–6, 49, 76, 79, 80, 84–5, 88–9, 160, 164–7, 200, 202–3, 204, 205; affectual 107; and artifacts 172–4, 182–5; and biological determinism 5; and choice theory 103, 111, 112; and collective memory 176; consequences of 97–8; and constitution 73–9; constraints on 135–6; effects of 118; and frame of reference 112–14, 120–3, 129–35, 137, 143; human models of 89; immaterial patterns of 172; institutionalized 13; intersubjective understanding of 102–3, 124–38, 140, 159, 191–9, 204; and knowledge 100–38; and materiality 4; and mechanical–Euclidean space 186–8; and mobility 169; norm-oriented model of 102–3, 117–24, 137, 138, 140, 186–90; objective aspects of 104–5, 106, 108, 109, 118, 202, 205–6; orientation 102, 106–7, 129–35, 160, 174, 178, 180; physical–material aspects of 140, 143, 200–1; and the physical world 183–5; and Popper's theory 30–5, 36, 97;

the purposive–rational model of 18, 102, 103–17, 123, 128, 138, 140, 181; rational 89, 179, 182; rational choice theory of 102, 103, 109–14, 117, 137; scientific theories of 39; situation 116–17; social 12, 92, 124–38, 147–50, 155, 156, 157, 183–5, 189–90; and society 13–14, 15, 18, 19, 189; sociological theories of 99, 100–38, 200, 206; and space 139–67, 186; and spatial differentiation 179; and spatial problems 142, 158; subjective aspects of 104–5, 106, 108, 109, 110, 118, 127, 204, 205–6; and symbolic places 174–5, 177–8; traditional 107; value-rational 107; Weber's three categories of 183–5

action theory 3, 6, 7–8, 13, 35–6, 100–38, 148, 154, 164, 167, 179, 181, 186, 187, 201, 203; and behaviourism 8, 9–20; logical 103–17, 179, 186; non-logical 105–6, 180; research 19, 101, 147–8; and social geography 91, 100, 139–67, 168–78, 179, 183, 185, 191; and sociology 168, 202; and spatial dimensions of the physical world 168–99; and structural-functionalism 117–18; voluntaristic 119